AUSTRALIAN ENVIRONMENTS

The Brigalow Belt

Cover illustration of Central Greater Glider: Artist: Peter Schouten (with permission)

ISBN: 978-1-7641194-3-6 (Paperback)

First edition: June 2025
A CONSERVATION CLUB PUBLICATION
www.conservationclub.com.au

NATIONAL LIBRARY OF AUSTRALIA

A catalogue record for this book is available from the National Library of Australia

In the spirit of reconciliation, Conservation Club acknowledges the Traditional Custodians of Country throughout Australia and their connection to land, sea, and community. We pay our respects to their Elders, past and present, and extend that respect to all Aboriginal and Torres Strait Islander peoples.

To my grandchildren: Samuel, Jameson, Harry, and Audrey

.. in the hope that they will enjoy sharing this journey of discovery ..

"And, above all, watch with glittering eyes the whole world around you because the greatest secrets are always hidden in the most unlikely places. Those who don't believe in magic will never find it."

Roald Dahl

CONTENTS:

REWARDS:

Conservation Club offers rewards (such as free downloads of colouring sheets and button badges) that can be earned by answering the 20 Question Quiz on page 148 and submitting one coloured-in image with the colours being truly representative of the species concerned. We encourage keen eyes, writing, recording, and reporting species sightings.

- **Buy one of our pocket notebooks! Designed to fit into the pocket of a child: at approximately A7 size, this blank notebook is an invaluable tool to have in order to record your observations of the environment you are exploring. It is really important to note the date, time of day, weather conditions and details of any sightings you may be lucky enough to experience! If the species is on the NCA, EPBC and/or IUCN Red List, please consider reporting your observations to an entity such as iNaturalist or Atlas of Living Australia! Citizen scientists are making great contributions to our understanding of these environments – you could be one of them!**
- **It is totally okay for you to guide your youngster and help them with this task (we suggest using the form at the back of the book as a guide before downloading the questions)!**

Email: conservationclubau@gmail.com for more information about how to apply for these rewards. The quiz questions are on the last page of this book. You can also join our mailing list to keep up to date with new releases.

Author's notes:

- Words and phrases shown within the book text in **BOLD** appear in the glossary/word lists at the back of this book.
- The whole point of the colouring pages is to encourage your young **conservationists** to engage with the book, tweak their observational skills, do a little bit of research, with the help of an adult if needed, and find out what colours are appropriate for the **species** concerned!
- Some of the words and language used may well be beyond what your young **conservationist** can understand, but it is suggested that you use the glossary to explain those words/phrases in simpler terms and at a level appropriate for your young wildlife warrior. They will ask for more information when they are ready!
- You can also choose to share the facts that you feel your young wildlife warriors are able to understand—you do not have to use them all! Just pick a few according to your child's age and comprehension skills.... You can always come back to the text later to expand their knowledge and vocabulary as they learn.

COLOURING RECOMMENDATIONS:

We recommend using colour pencils rather than texters or felt pens to ensure the information pages remain readable. Texters can bleed through the colouring side, affecting the readability of the fun facts!

We encourage adults and youngsters to find coloured images or photos online to guide the colouring. Familiarising both you and them will assist in being able to recognise the **species** if you sight them!

INTRODUCTION:

In these colouring pages, we focus on **The Brigalow Belt. The Brigalow Belt** is a wide band of acacia-wooded grassland that runs between the tropical rainforest of the coast and the **semi-arid** interior of Queensland and northern New South Wales, Australia. The Interim Biogeographic Regionalisation for Australia (IBRA) divides the Brigalow Belt into two IBRA regions, or bioregions, *Brigalow Belt North* (BBN) and *Brigalow Belt South* (BBS). The north and south Brigalow Belts are two of the 85 bioregions across Australia and the 15 bioregions in Queensland. Together, they form most of the Brigalow **tropical savannah** ecoregion.[2]

Brigalow Country is over 500,000 km2 in size, from Bowen, Queensland, in the north to Northern NSW in the south. It is recognised for its **conservation** values, **agriculture**, natural tourist attractions, mining and gas resources. For thousands of years, Brigalow Country has been home to many First Nations peoples, who are the Traditional Owners of various lands within the priority place.

Brigalow is a **species** of acacia (*Acacia harpophylla*), an ecological community and also the name of the wider bioregion. The **biodiversity** of Brigalow country is **diverse** and **unique**, with many **species** in the region found nowhere else in the world.

Brigalow country sits in a transition zone, between the coast to the east and the **arid** regions to the west, and from the **tropical** north to the more **temperate** south. This transition zone provides a range of different natural **habitats** from mountains to lakes and wetlands. These have a variety of vegetation types ranging from dry scrublands to forests and grasslands.

This large, **complex** strip of countryside covers an area of undulating to rugged slopes, consisting of ranges as well as plains of ancient sand and clay deposits, basalt and alluvium. The Mulga lands and the Brigalow Belt are where most of Queensland's land clearing is occurring. Much of the Brigalow woodland has been cleared or radically reduced to the extent that some wildlife, failing to thrive in the altered **environment**, has become extinct here, with a number of the remaining communities threatened or endangered. The clearance of brigalow and poplar box is ongoing as there are a number of nature reserves of which do protect the various types of **habitat** found in the Belt including brigalow and eucalyptus woodland, grassland, vine thicket, high peaks, sandstone gorges and wetlands however these tend to be located on the sandstone uplands rather than the fertile lowlands, where the brigalow woodlands are still vulnerable to clearance and are often limited to small areas of parkland.[2]

The grasslands of the region are also under threat from **introduced** pasture grasses such as buffelgrass and weeds such as Congress weed. One particular threat comes from alterations to natural flow patterns caused by the addition of dams and weirs, which impact the riverine and floodplain plant and animal **species.**[2]

Some of the **species** featured in this book are **critically endangered**, **endangered**, or **vulnerable**; the illustrations include this information where appropriate and available.

The Brigalow Belt bioregion is experiencing **significant** wildlife challenges, with 147 **species** and 100 ecological communities listed as threatened. Approximately 80% of the Brigalow Belt south bioregion is **grazed**, and there have been **declines** in certain bird **populations**, particularly ground-nesting, ground-feeding **insectivores** and grassland birds. There are 21 threatened ecological communities in Brigalow Country, roughly 20% of all listed communities.

WILDLIFE IN THE BRIGALOW BELT[3]

Current statistical data identifies:

- a total of 1582 animals in the Brigalow Belt, 1534 of which are **native** and 48 **introduced species**. Of this number, 156 are considered to be **rare** or **threatened species**.
- a total of 47 **bacteria**
- a total of 145 chromists (**microorganisms**)
- a total of 970 **fungi**
- a total of 7113 plants, of which 6059 are **native** and 1049 **introduced**. Of this number, 228 are considered to be **rare** or **threatened species**. That does leave five unaccounted for!
- a total of 4 **protozoans**

Note: Wildlife statistics are based on information that has been submitted to the DETSI WildNet database and converted to a 10km² grid. The grid information has been intersected with the mapping polygons to determine the **species** lists.

Information from WildNet can also be accessed via Wildlife Online and Wetland*Maps*.

Rare or threatened includes **species** listed as extinct, extinct in the wild, critically endangered, endangered, vulnerable or conservation dependent under either the Nature Conservation Act or Environment Protection and Biodiversity Conservation Act.

Protected areas:

<u>A little more than 2% of the Brigalow Belt lies within national parks and other protected areas</u>.

Further information about National Parks is given on pages **5** and **6**.

Australia is the sixth largest country on Earth and the only island/country that covers an entire **continent**! Because of its **isolation** from other **continents**, many of Australia's animal and plant **species** are found nowhere else. Australia also has a wide variety of **environments**, many of which are threatened for various reasons, including climate change, **extreme weather events**, and human impacts.

DID YOU KNOW THAT around 80% of Australia's **native species** are **endemic,** and many are likely to become **extinct** within our lifetimes? The Conservation Club AU focuses on raising awareness of **species** and **environments** in our 'backyards,' which we may not know. Based on **fossil** records, the natural **extinction** rate is approximately one **species** per one million **species** yearly. Scientists estimate that around 8.7 million **species** of plants and animals currently live on Earth. They are racing against time to catalogue Earth's **biodiversity** as **extinctions** continue. The Australian **continent** has seen more **mammal extinctions** than any other, along with **significant** losses of frogs, lizards, fish, insects, and plants. Many of these **extinctions** are recent. In Australia, there have been 21 likely **extinctions** of animals and four of plants, averaging about four per **decade**. Out of the 23 **extinctions** for which causes have been identified, 17 (73%) were primarily due to **invasive species**. Focusing just on animals, 17 out of 19 (89%) **extinctions** were mainly caused by **invasive species**. The four plant **extinctions** are primarily attributed to **habitat destruction** or **degradation**. The **primary culprits** in recent **extinctions** include the ***chytrid*** **fungus** (affecting **amphibians**)**, which has wiped out six **species**, and the wolf snake, which is responsible for the loss of four **species**. Other **significant** impacts include cats, foxes, black rats, brown trout, and a ***trypanosome*** blood **parasite**. Although **habitat** loss is often **cited** as the **primary** cause of **extinctions**, this does not hold true for recent **extinctions** in Australia. Earlier conclusions linked many plant **extinctions** to **habitat** loss. Still, increased efforts in searching have led to the rediscovery of many of these plants. The current data indicates that **invasive species** have been the leading cause of most **extinctions** in Australia since **European settlement.**

LET'S EXPLORE!

Kids in the 8 to 14-year-old age groups are generally interested in spending time with their families exploring the great outdoors on weekend walks, camping getaways, etc. What a great way to spend time with your kids! It is so much better than burning up hours using electronic media and becoming **sedentary**, both of which can lead to behavioural problems and issues with sleep!

Kids also love to be part of a club and start collecting things they might be interested in, such as pebbles and shells. In my opinion, our kids are generally far more aware of **conservation** and **environmental** issues than in previous generations. This, therefore, presents an opportunity to educate, to get them interested, active, **motivated,** and involved in the **environment** and the **conservation** of **species** so that they will, hopefully, be able to work to preserve these animals so that they can show or tell their own children about them in years to come.

PLEASE…. GET OUT THERE and encourage these young and curious minds to engage with their **local environments** and the **critters** that can be found in them!

NATIONAL & MARINE PARKS:

Here are lists of some of the national and marine parks in New South Wales and Queensland, which include Brigalow Belt environments to get you started!

These lists do not cover all the national or marine parks and do not include any nature reserves, conservation parks, etcetera as individual listings.

NEW SOUTH WALES NATIONAL PARKS IN THE BRIGALOW BELT

This list may not cover all areas of Brigalow, and some areas straddle the borders of the Brigalow Belt South bioregion. The area around Lighting Ridge has **remnant** areas of Brigalow; although these are not attached to the BRB South greater area, they still form a part of it!

- Barayamal National Park
- Breelong National Park
- Bullala National Park
- Burral Yurrul National Park
- Coolah Tops National Park
- Couradda National Park
- Crooked Creek National Park
- Dowe National Park
- Drillwarrina National Park
- Dthinna Dthinnawan National Park
- Garrawilla National Park
- Goonoo National Park
- Gunyerwarildi National Park
- Gwydir River National Park
- Kwiambal National Park
- Moema National Park
- Magriguy National Park
- Mount Kaputar National Park
- Pilliga National Park
- Pilliga West National Park
- Rocky Glen National Park
- Sappa Bulga National Park
- Somerton National Park
- Timmallallie National Park
- Tinkrameanah National Park
- Warialda National Park
- Warrumbungle National Park
- Yarrigan National Park

QUEENSLAND NATIONAL & MARINE PARKS IN THE BRIGALOW BELT

For more information in Queensland, go to https://wetlandinfo.des.qld.gov.au/wetlands/facts-maps/bioregion-brigalow-belt-brb/

- Albinia National Park
- Alton National Park
- Auburn River National Park
- Baga National Park
- Bania National Park
- Beeron National Park
- Bendidee National Park
- Blackdown Tableland National Park
- Blackwood National Park
- Bowling Green Bay National Park
- Broad Sound Islands National Park
- Bunya Mountains National Park
- Cania Gorge National Park
- Cape Upstart National Park
- Capricorn Coast National Park
- Carnarvon National Park
- Castle Tower National Park
- Chesterton Range National Park
- Culgoa Floodplain National Park
- Curtis Island National Park
- Dan Dan National Park
- Dawes National Park
- Dipperu National Park (scientific)
- Epping Forest National Park (scientific)
- Erringibba National Park
- Eungella National Park
- Expedition National Park
- Expedition (Limited Depth) National Park
- Gloucester Island National Park
- Goodedulla National Park
- Great Barrier Reef Coast Marine Park
- Homevale National Park
- Humboldt National Park
- Isla Gorge National Park
- Junee National Park
- Kroombit Tops National Park
- Main Range National Park
- Mazeppa National Park
- Minerva Hills National Park
- Mount Abbot National Park (scientific)
- Mount Aberdeen National Park
- Mount Archer National Park
- Mount Etna Caves National Park
- Mount O'Connell National Park
- Nairana National Park
- Narrien Range National Park
- Nuga Nuga National Park
- Palmgrove National Park (scientific)
- Paluma Range National Park
- Peak Range National Park
- Pinnacles National Park
- Precipice National Park
- Redcliffe Tableland National Park
- Rundle Range National Park
- Snake Range National Park
- Southwood National Park
- Taunton National Park (scientific)
- Tregole National Park
- Wietalaba National Park
- Wondul Range National Park

Some background information about this book:

The **species** in this book occur in the Brigalow Belt. Many (but not all) are on the *'Rare or threatened animals of Brigalow Belt North (BBN) IBRA bioregion list',* and/or the *'Threatened* ***species*** *found in Brigalow Belt South (BBS) IBRA'* lists. Some of the **species** illustrated are either **Critically Endangered**, **Endangered**, or **Vulnerable** on the *NCA* and *EPBC (SPRAT)* lists.

On the IUCN Red List.[1]*, despite this being extensive, more than 169,420* ***species*** *have been assessed at the time of writing this book. However, of the goal of 260,000* ***species****, there are still 90,580* ***species*** *remaining to be evaluated, an incredible achievement, but, for some* ***species****, due to the lack of up-to-date scientific data, this list is, at this point in time, not complete. The list is constantly updated, with a goal to reassess 142,000* ***species****. More information can be found at https://www.iucnredlist.org/*

The following tables give information about how the ***species*** *are listed with NCA, EPBC (within Australia), and IUCN lists, within what categories, and a breakdown of the categories currently available for assessment.*

Younger readers may be unable to navigate this information themselves, so it is provided here to enable parents or caregivers to relay it to those young, curious minds!

I firmly believe that the age groups this book is aimed at (8 to 16 years) are curious and generally more informed than previous generations. They are aware that even one single voice can make a difference. This is evidenced by the number of young people taking up a cause. They recognise that, even one voice, one person, can make a difference, as evidenced by the number of young people taking up a cause.

The book is likely to appeal to a much broader audience than just the age group mentioned, as the illustrations and information are also appropriate for adults. Most importantly, we all need to be much more aware of our surroundings and report our observations to an entity such as SPRAT.

I sincerely hope that you, the Wildlife Warriors for the Brigalow Belt environment, will enjoy this book as much as I have enjoyed writing and illustrating it!

Elysia Rainwood

NCA – Status under the Nature Conservation Act 1992 (QLD legislation) [21]

EX	Extinct
PE	Extinct in the wild
CR	Critically endangered
E	Endangered
V	Vulnerable
NT	Near threatened
C	Least concern
SL	Special least concern
I	International
P	Prohibited

EPBC – Status under the Environment Protection and Biodiversity Conservation Act 1999 [2]

EX	Extinct
XW	Extinct in the wild
CE	Critically endangered
E	Endangered
V	Vulnerable
CD	Conservation dependent

IUCN – Status on the IUCN Red List [1]

EX	Extinct
EW	Extinct in the wild
CR	Critically endangered
EN	Endangered
VU	Vulnerable
NT	Near threatened
LC	Least concern
DD	Data deficient
NE	Not evaluated

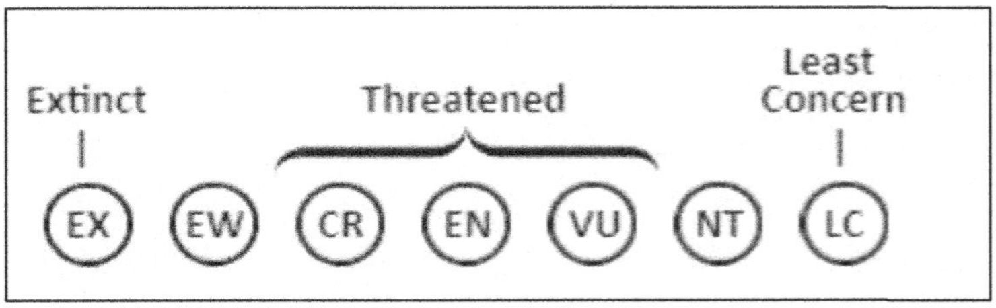

Where the species are on the **IUCN Red List,** the category is outlined in bold and grey in colour!

<u>**IUCN categories explained:**</u> [1]

CE = Critically Endangered:....... Facing an extremely high risk of extinction in the wild.

EN = Endangered:...................... There is a very high risk of extinction due to rapid population declines of 50 to more than 70% over the previous 10 years, a current population size of fewer than 250 individuals.

V = Vulnerable:.......................... Being threatened with extinction unless the circumstances that are threatening its survival and reproduction improve.

- Sometimes the species on the IUCN Red List have not been listed simply because there has not been sufficient research to assess.
- In these cases, they are regarded as DD = Data Deficient **OR** NE = Not Evaluated.
- That does not mean that they are not critically endangered, endangered, vulnerable, threatened or any other classification... It just means that they have not been assessed!

Where on earth is the Brigalow Belt found?

The term "brigalow" refers to a particular species of acacia and also to the name of the wider bioregion.
Here are some maps (Queensland and New South Wales, Australia) which help to delineate further the areas of the Brigalow Belt, both north and south.

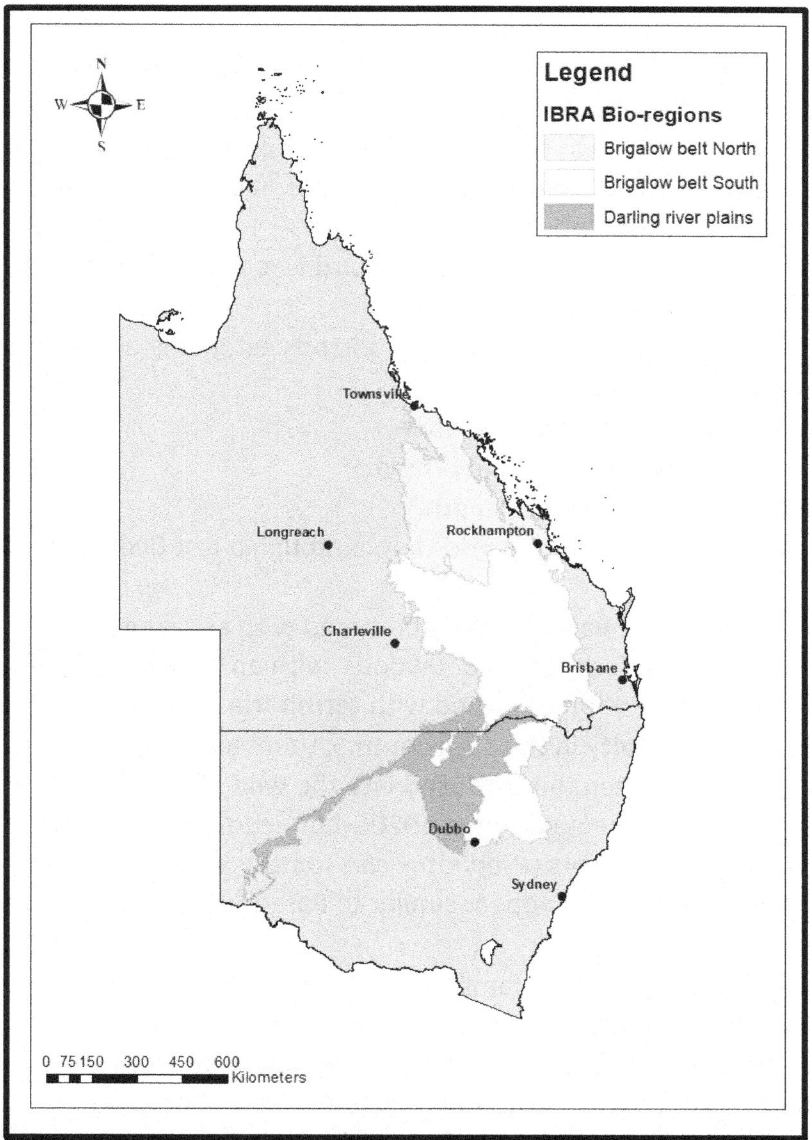

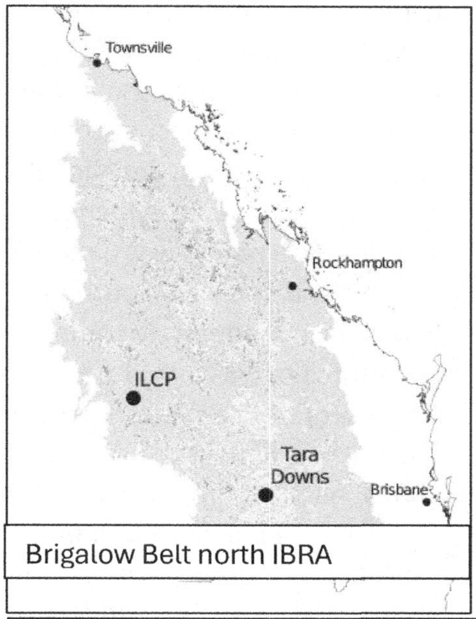

Brigalow Belt north IBRA

Brigalow Belt south IBRA

Paradise parrot **EXTINCT**

Psephotellus pulcherrimus

- Sadly, this extraordinarily colourful bird, even by parrot standards, is the only mainland Australian bird recorded as extinct by the International Union for the Conservation of Nature (IUCN).
- The extinction of the Paradise parrot is likely to have resulted mainly from a combination of new pastoralist practices which **introduced** the burning by **graziers** to encourage fodder for their stock resulting in the loss of seasonal foods (at the critical time when the grass was in seed), **overgrazing** and land clearing, in combination with the effects of a severe drought in 1902, so that little amounts of seed were produced for several years. Hunting by bird collectors and predation by **introduced mammals** like cats and dogs may also be reasons.[4]
- Its premature disappearance almost a century ago, meanwhile, remains prophetic today when it comes to how best to protect other threatened Australian avian **species**.
- That the Paradise parrot – *Psephotellus pulcherimus* – was already on the verge of extinction by 1900 in its **habitat** on the Darling Downs[5] in the Queensland **colony** speaks volumes about the dramatic **environmental** impact of colonisation on **native** grassy woodlands.
- The last live bird was seen in 1927. Extensive searches since the last live bird was seen have failed to produce any reliable evidence of its continued existence.
- The Paradise parrot was a medium-sized parrot **native** to the grassy woodlands extending across the Queensland and New South Wales border area of eastern Australia.[4]
- It was once moderately common within its somewhat **restricted range**.
- The **plumage** was a mixture of turquoise, aqua, scarlet, black and brown.
- The parrots were small, measuring approximately 27-30cm in length.
- The tail was almost the same length as the body; unusual for a bird that, although a fast flyer, spent almost all of its time on the ground.[4]
- The Paradise parrot mainly **inhabited** undulating river valleys, lightly timbered with eucalypt woodlands or open forests, often dominated by ironbarks and bloodwoods, with an **understorey** of **annual** and **perennial native** grasses; these areas were often dotted with **termitaria**.
- Though many Paradise parrots were taken into **captivity** in the 19th century, there are no current captive **populations** of this **species**, and none have been re**introduced** into the wild. There was a proposal to begin **captive breeding** of the **species** for release in the 1920s, but action occurred too late to save the **species**. Hybrid golden-shouldered parrots (*Psephotus chrysopterygius)* and Mulga parrot (*P. varius)*, which occur only in **captivity,** are said to appear similar to Paradise parrots but should not be mistaken for them.
- They were usually recorded in pairs or small **flocks**, probably family groups, of up to seven or eight, making their nests in hollowed-out termite mounds, often at or near ground level, and feeding, so far as is known, almost exclusively on grass seeds. Some observers state that **termitariums** were occupied by the same breeding pair in successive years. Less certain is confirmation of holes at riverbanks being tunnelled out by this **species**. A recorded **range** of one pair over a month was **restricted** to an area of two hectares. The nest sites were occasionally reported at **termitariums** previously occupied by the kingfisher *Todiramphus macleayii* or in **burrows** of a kookaburra *Dacelo leachii*, or at the base of a tree.[4]

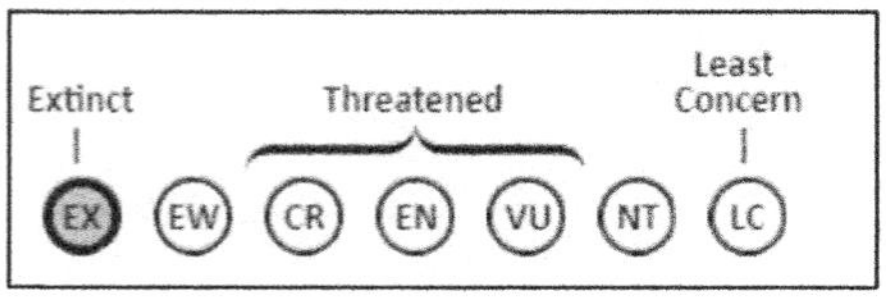

[1]

Western quoll (chuditch) ....LOCALLY EXTINCT

Dasyurus geoffroii geoffroii

- The western quoll (chuditch) is currently not found in the Brigalow Belt, though it was previously a part of its **ecosystem**. The Brigalow Belt is a region facing ecological challenges, including land clearing, **agricultural** expansion, and **introduced species**, which have contributed to the **decline** of **native** wildlife, including the western quoll.
- Western Quolls were once present in a wide variety of **habitats** across nearly 70% of the Australian mainland. Following **European settlement**, their **range** contracted dramatically. They are now found only in the south-western corner of Western Australia, and even within this region, their **distribution** is patchy.[6]
- The western quoll is a large **carnivorous marsupial recognised** by its body pattern of white spots on red-brown fur. Their 30cm long tail ends in a black bush from halfway down to the tip. They grow to a body length of 46cm and have large eyes and pointed ears.
- The western quoll was once widespread across much of Australia, but its **range** has shrunk **significantly**, primarily to the far southwestern corner of Western Australia.
- The Brigalow Belt north bioregion, which includes parts of Queensland, has a notable list of threatened **species**, including the western quoll. However, the **species** is **locally extinct** from this bioregion.
- The western quoll's **decline** is attributed to various factors, including:
    - **Predation** by **feral** cats and foxes is a serious threat to the western quoll, and their introduction has had a devastating impact.
    - **Habitat** modification is likely to have contributed to the **decline** of western quolls through the combined influences of land clearing, inappropriate **fire regimens** and **grazing** by both stock and **feral herbivores**.
    - **Illegal** shooting and **poisoning** are also likely to be partly responsible for the disappearance of western quolls from heavily populated and **agricultural** areas.
- Western Quolls are **solitary** animals with very large **home ranges**, reflecting their **carnivorous** feeding habits. Their diet is made up of large **invertebrates** and a variety of **reptiles**, birds and **mammals** (up to the size of bandicoots and parrots). Animals are primarily **nocturnal**, hunting at night and sheltering in hollow logs or **burrows** during the day.[7] Young are born between May and September and are independent at about six months of age. Individuals become sexually **mature** at one year of age and usually do not live beyond three years.
- **Conservation** efforts: Organisations like the Australian Wildlife Conservancy are actively involved in protecting and reintroducing western quolls to suitable **habitats**.

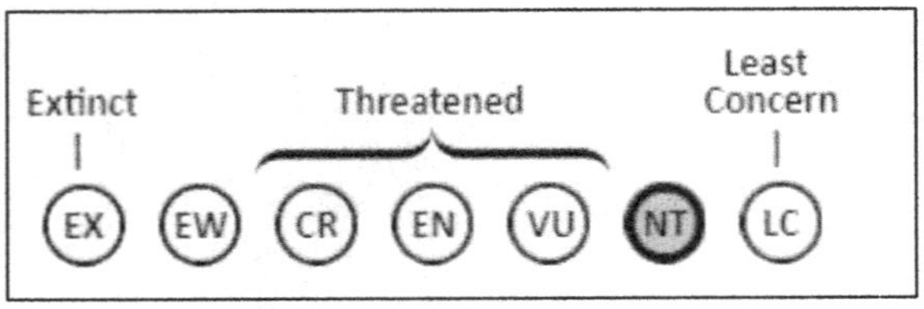

[1]

QLD NCA List = EX = Locally extinct
NCA = V = Vulnerable
EPBC List = V = Vulnerable

Regent honeyeater

Anthochaera phrygia

NATIVE TO AUSTRALIA

- The scientific name comes from the ancient Greek *Anthos*, "flower, bloom," and *khairo*, "enjoy," and the **specific species** name from Phrygius, a reference to the people of Phrygia (now Asian Turkey) who were skilled in the use of gold in embroidery.
- The regent honeyeater's diet consists mainly of **nectar** from Eucalyptus and mistletoe **species**. Sometimes they feed on insects and their **honeydew** as well as **native** and **cultivated** fruit.
- Only about 250-300 birds are **estimated** to remain in the wild.
- They are great **mimics** of other birds' songs but have a **unique** and **complex** song of their own. Research suggests that part of the reason for the **rapid decline** in **population** is that the young are struggling to learn the mating calls from the adult birds, which may also reduce their ability to find a mate and to breed successfully.
- **Captive breeding** programs are in place in New South Wales.
- The Pilliga woodlands are known to be an important area for breeding and foraging for regent honeyeaters.

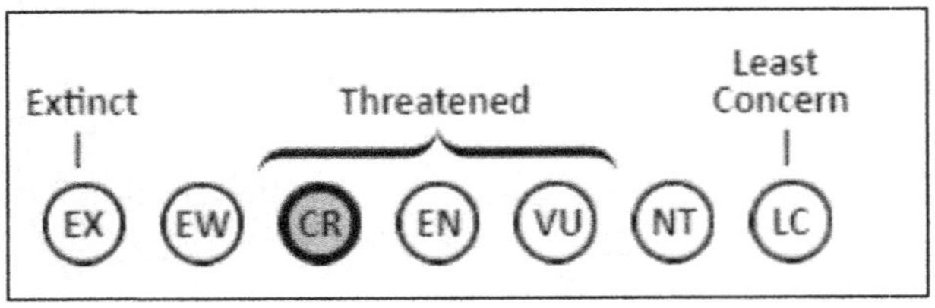

[1]

NCA = CR = Critically endangered

EPBC = CE = Critically endangered

To learn more about these birds, watch the video from Taronga Western Plains Zoo: https://www.youtube.com/watch?v=Y4680qGgUrE

Coxen's Fig Parrot

Cyclopsitta diophthalma coxeni

NATIVE TO AUSTRALIA

- Coxen's fig parrot is one of Australia's rarest and least known birds.
- The average total length is about 14 - 16cm, and it is the smallest parrot in Australia.
- Coxen's fig parrot is **cryptic** and extremely difficult to see in its **habitat,** so it may be more common than the number of sightings suggests.
- It has been recorded on just over 200 occasions since it was described by Gould in 1867.
- Although signs of **excavated** nests have been found, no active nests have been recorded. Coxen's fig parrot **excavates** its nesting hollow in the decaying wood of dead or living forest trees, as with other **subspecies** of double-eyed fig parrots.
- Coxen's fig parrot feeds mainly on figs and other fleshy fruits. It has also been recorded feeding on the **nectar** of silky oak (*Grevillea robusta)*. Different **subspecies** are also known to eat insect **larvae**.
- Confirmed or **credible** sighting reports continue to be made in both **range** states, including about 30 records in north-east New South Wales since 1970 and twice this number in south-east Queensland over the last **decade** alone. Nevertheless, accurate **predictions** about **population** size are currently not possible.
- Threats include clearing of **subtropical** rainforest for **agriculture**, especially winter **foraging habitat** in coastal lowlands and **degradation** by logging of rainforests.
- Coxen's fig parrot has very **distinct** markings, with no red above the beak. It should not be confused with Marshall's fig parrot or Macleay's fig parrot, which are VERY similar!

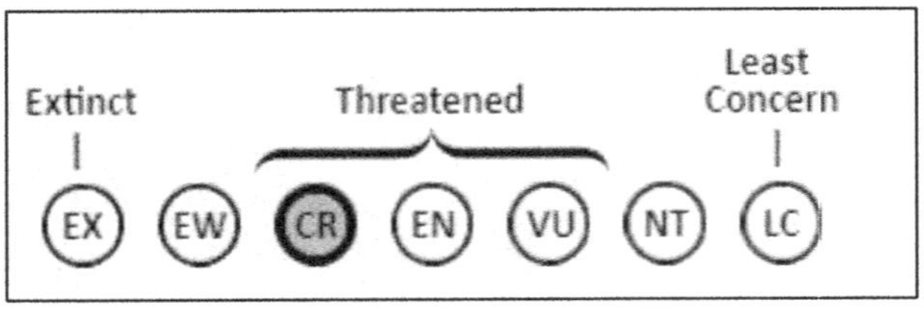

[1]

NCA List = CR = Critically Endangered

EPBC List = CE = Critically Endangered

White-throated snapping turtle

Elseya albagula

ENDEMIC TO AUSTRALIA

- The white-throated snapping turtle is slow-growing and is thought to reach sexual **maturity** around 15-20 years old in the wild. Adult females nest once yearly, laying a single **clutch** of eggs, with 6-10 eggs per **clutch**.
- They have powerful jaws and have an average **bite force** of 209N (Newtons) of pressure. They may not have the same force that you do when you bite with your **molars** (1300N), but they have enough force to do some real damage.
- Common snapping turtles can hold their breath for at least 30-40 minutes before taking a quick breath!
- It is the largest **species** of snapping turtle (*Elseya sp.)* still in existence, with **carapace** lengths of up to 42cm in females and up to 30 cm in males (adults in the Mary River are smaller on average than their respective **counterparts** in the Burnett and Fitzroy rivers). Snapping turtles are slow on land, travelling under 2 km/hour on average. In water, they travel 16-19km/hour for long distances and can reach 35km/hour at maximum speed.

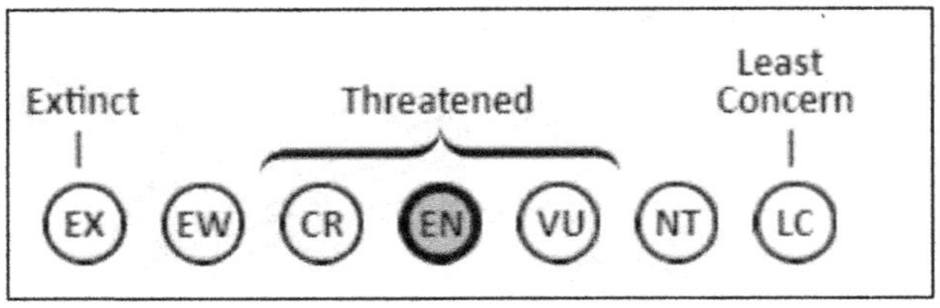

[1]

NCA List = CR = Endangered

EPBC List = CE = Critically Endangered

Plains wanderer

Pedionomus torquatus

- These shy little birds are found on the bare saltbush plains or in the grasslands and herb fields of northern Victoria and outback New South Wales, stretching into Queensland.
- Plains wanderers are so shy and **secretive** that even scientists rarely see them. Researchers estimate there are fewer than a thousand of these **unique** birds left in the wild.
- The best evidence of their presence is taken from automatic recording devices sensitive enough to pick up their gentle 'ooo' calls at **dusk** and dawn. These recording devices have recently become more silent, indicating that the Plains-wanderer is struggling to survive.[8]
- The Plains-wanderer is an avian oddball, a mysterious little grassland specialist. There are no other birds like them in the world. Elusive and shy, these birds are most often encountered spotlighting at night, startled from their **roost** on the ground. They crane up on tiptoe when alerted to danger, then crouch down to make a dash for cover.
- Standing about 12-15cm tall with a wingspan of 28-36cm and weighing less than 100 grams, Plains-wanderers are small fawn-coloured quail-like ground-dwelling birds. Their dappled feathers include white and blackish marks, with spots and streaks on the head and neck. The larger female is easily distinguished by her **prominent**, white-spotted diamante collar above a rich **rufous** breast patch.
- Plains wanderers live in semi-arid, lowland **native** grasslands. Their nests are shallow, grass-lined scrapes that are scratched into the ground. They **forage** during the day for a wide variety of seeds and ground-dwelling insects and do not require regular access to water bodies; instead, they fulfil their hydration requirements through the food they **consume** and via pecking dew and raindrops from leaves. Each bird is **solitary** and has an average home **range** of around 12 hectares. Males and females with overlapping **ranges** form breeding pairs, with the larger females **defending** their territories and mating with several birds in a season, while the males **incubate** eggs and raise the young.[9] In favourable conditions, females can lay multiple **clutches** of three to five eggs per year.
- **Predation** is a danger that is ever-present from dingoes, foxes, **feral** cats, and falcons - the **species**' ground-nesting habits, poor flying ability, and tendency to run rather than fly from **predators** make it easy **prey**.
- Most **significant** is the loss of **habitat** to farms. Over and **undergrazing** by **introduced** and **native herbivores** and invasion by weeds significantly decrease **habitat** quality. Other **threats** include the impacts of altered **fire regimens**.
- The **survival** of the Plains wanderer will depend upon collaboration between zoos, farmers, scientists, wildlife organisations, and broader climate change mitigation action.

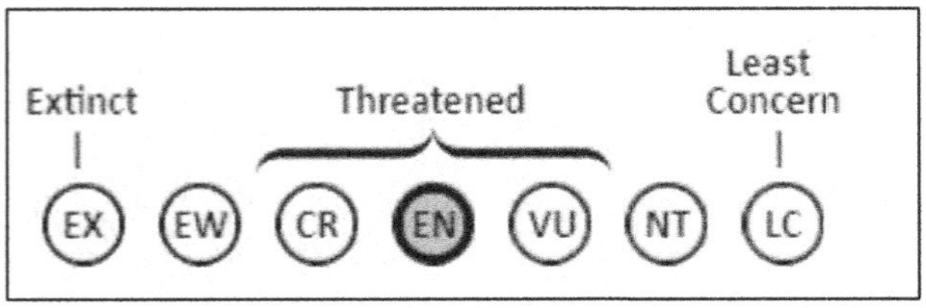

[1]

NCA List = CR = Endangered

EPBC List = CE = Critically Endangered

Bulloak jewel

Hypochrysops piceatus

- The bulloak jewel butterfly is thought to be the second rarest butterfly in Australia.
- First discovered in the late 1960s. The bulloak jewel butterfly lives on the bulloak (*Allocasuarina luehmannii*) trees. While the tree from which the caterpillars draw their name is not rare in itself, the bulloak woodland **ecosystem** is listed as an endangered ecological community.[10]
- Its **distribution** is mainly concentrated near Leyburn, about 70km south of Toowoomba, Queensland, in the area of land occupied for thousands of years by the people of the Githabul dialect of the Bundjalung people. Known as the Ellangowan Nature Refuge, this strip is the stronghold of this tiny, rare insect.[10] The main breeding sites are near Ellangowan Nature Refuge near Leyburn and the Bendidee National Park/State Forest area.
- The butterfly looks like a jewel! Underneath the wings is a greyish background, with little golden, silver, green and bright red spots all arranged in the most extraordinary way.
- Butterflies rely on **mature** (100 to 200-year-old) bulloak for **larval** feeding and shelter and a **specific** ant **species** (*Anonychomyrma inclinata*) for protection.
- The ants pick up the caterpillars, carry them into their nests or into the hollow twigs, and they actually look after them. At night, the ants herd the caterpillars out to bulloak leaves — the only food that those little caterpillars will eat. Then, in the hot daytime, the ants will steer the caterpillar back to cool safety, in hollows or lightning scars in large, old bulloak trees. The ants protect the caterpillar from **parasites** such as flies, which like to lay their eggs into the caterpillar, so their own babies **hatch** out within a smorgasbord of fresh insect guts.[10]
- The caterpillars produce an attractive secretion, which is made up of honeydew, which attracts the ants. The ants like sweet **nectar**, like gum or mistletoe blossom. Incidentally, this is the same food the adult bulloak jewel butterfly will require to survive.[10]
- The butterflies are so dependent on this relationship that the ants must be present for them to **reproduce** successfully. In fact, she [the adult butterfly] won't lay her eggs unless there's evidence of the ants on that tree.[10]
- There are a total of seven **instars** from the beginning to the end of development.
- The bulloak jewel butterfly faces several threats, including **habitat** loss and **destruction**, fire, climate change, and the presence of a particular ant **species** that the butterfly relies on for **survival**. The butterfly's **habitat** is being lost due to land clearing and natural events like drought and fire. Additionally, climate change and increasingly dry conditions are impacting the butterfly's ability to survive.

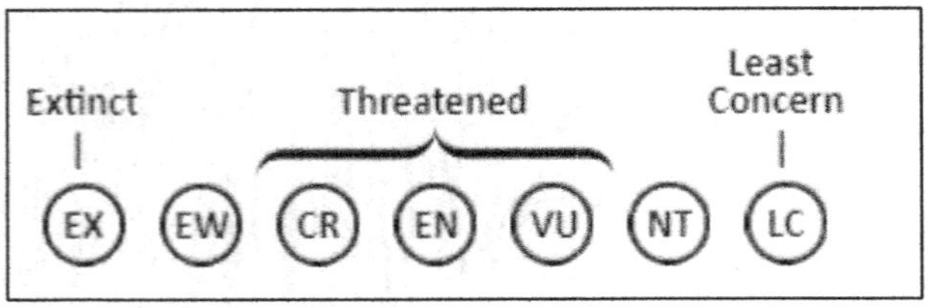

[1]

NCA List = CR = Endangered

EPBC List = CE = Critically Endangered

Gulbaru gecko

Phyllurus gulbaru

- Ancient and spectacular, this endangered gecko has one of the smallest **distribution**s of any Australian animal.
- The Gulbaru gecko has a very small and patchy **distribution** into two areas, even within the suitable, severely **fragmented habitat**. This means there is little **migration** between **populations**. The larger population of Gulbaru geckos is protected in Mt Cataract Forest Reserve and Paluma Range National Park in the Townsville area, just south of the UNESCO Wet Tropical World Heritage site boundary, and only a small fraction of the suitable **habitat** for this **species** is protected. The total area in which the Gulbaru gecko **species** lives is only 23km^2.[11]
- This large gecko, growing to 18 cm, is **restricted** to rocky rainforest, generally dominated by Hoop Pines (*Araucaria cunninghamii*). It hides in rock cracks during the day and emerges at night to hunt **invertebrates** on the rock surfaces. It is slow-moving and highly **camouflaged**.
- These geckos are notable for their highly effective **camouflage,** boosted by **tubercles** covering their bodies.
- Gulbaru geckos can **regenerate** their tails.
- Females lay two eggs, which develop slowly.
- Australia is a **global** centre of gecko **diversity**, with a remarkable 140 **species** at the last count. Australia's geckos fall into three families: *Diplodactylidae, Gekkonidae* and *Carphodactylidae*. The last of these, *Carphodactylidae*, is a **uniquely** Australian group, the only lizard family **endemic** to Australia. Among the **genera** within *Carphodactylidae* is *phyllurus*, which means "leaf tail". This family of Australian geckos is ancient. They diverged from all other living **reptiles** more than 60 million years ago.[11]
- There are nine **species**, **restricted** to coastal eastern Australia. The most southerly is found in the sandstones of the Sydney region. The remaining eight **species** are all **restricted** to tiny **distribution**s along the Queensland coast, in many cases to a single mountain or range. All these **species** are found in rocky rainforest. No two **species** are found in the same patch, and their distribution reflects the gradual contraction and fragmentation of rainforest in eastern Australia into small, **isolated** pockets over millions of years.
- They are rainforest dependent and can be found in gullies near running water.
- The **primary** threat is unmanaged burning, particularly late dry season fires that encroach into the rainforest from nearby open forest and pastoral areas. Fires are a natural part of the landscape in this region, but intense burning can chip away at the rainforest edge. **Invasive** grasses growing at the rainforest boundary provide a thick, highly flammable fuel load that can exacerbate these effects.
- Climate change, for example, could potentially threaten the species if it leads to drier conditions and greater fire potential. An unlikely but obvious direct threat is quarrying, an activity that does not occur within the **distribution** but does occur nearby in the region.

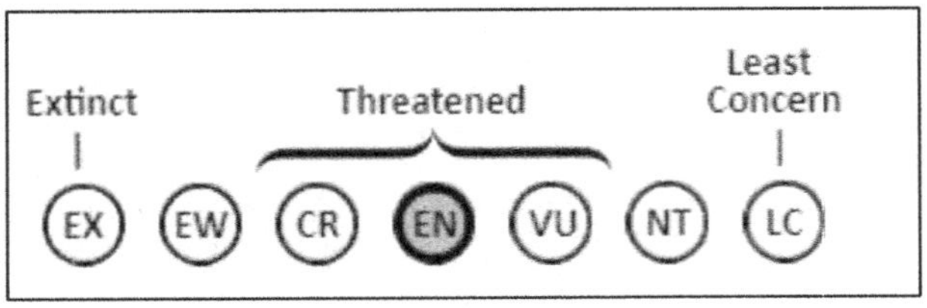

[1]

NCA List = CR = Endangered

EPBC List = CE = Critically Endangered

Kroombit treefrog

Litoria kroombitensis

- The Kroombit treefrog is a small frog **species endemic** to Australia.
- These treefrogs, which are hard to find and have little documented history, were first described formally in 2013 and are only known to live in the Kroombit Tops National Park, southwest of Gladstone.
- The **species** grows to not more than about 45mm in length (snout to vent length), with females being on average some 25% longer than males. Colouration is mostly green or greenish-brown with orange on the thighs and thin gold lines extending over the eyes and **tympana** to the front limbs.
- The mating call is a short whine followed by one or two chirps.
- The **species** is a stream breeder, laying clumps of 100–300 eggs embedded in jelly which are wrapped around submerged twigs in still or slow-flowing water.
- The **species** is found in the Central Highlands Region of Queensland, where it is **restricted** to the headwaters of five streams rising on the eastern side of Kroombit Tops, an ecologically **isolated** plateau covered with **temperate** wet forests, with an altitudinal **range** of 550–900m, that is surrounded by **subtropical** woodlands.[2][1]
- Although the Kroombit Tops National Park protects the species, threats to the population include the fungal **disease chytridiomycosis**; **habitat degradation** and water fouling caused by the activities of **feral** cattle, horses and pigs; as well as the impact of climate change, such as increased frequency and severity of bushfires.

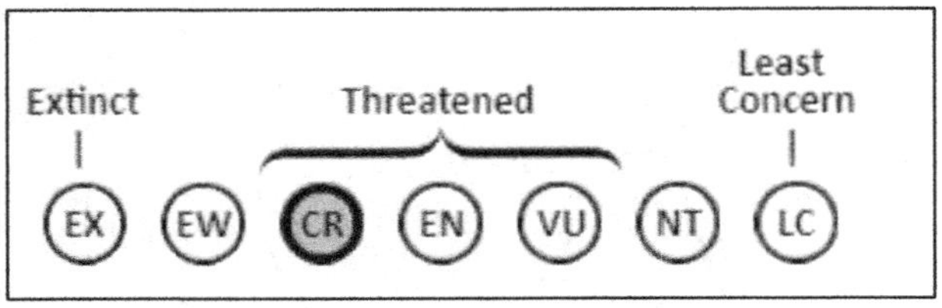

[1]

NCA List = CR = Endangered

EPBC List = CE = Critically Endangered

Jagara hairy crayfish

Euastacus jagara

- The Jagara hairy crayfish is only known from six **specimens** collected in the Mistake Mountains located within the Goomburra State Forest, about 80km southwest of Brisbane, Queensland. The Mistake Mountains are approximately 1027m above sea level.
- The Jagara hairy crayfish is **restricted** to the higher **altitude** rainforest community of Main Range National Park. The streams where it lives are surrounded by rainforest and drain into the Brisbane River.
- The Jagara hairy crayfish is a dark green or green-blue to orange-brown **dorsally**, with a bluish tint developing **ventrolaterally**. The legs are light blue or blue-brown, and the chelae have a distinct mottled pattern, with a greenish tint highlighted with a brighter blue. The larger **mesial carpal** spines are generally red with yellow tips.
- The **species** was recorded from four headwater streams, including the type **locality** (Flaggy Creek) and three branches of Dalrymple Creek, situated further south. Initially, it was suggested that the **species** may occur further south than the type **locality**, and the recent Australian Crayfish Project surveys confirmed this. Three of the sites **inhabited** are situated at the southern end of the **species ' range** on the opposite (western) side of the Great Dividing Range, in the Murray-Darling Basin (MDB) **drainage** division of inland Australia. The occurrence of the Jagara hairy crayfish in both the Northeastern Coastal Drainage (NEC) division and the MDB is noteworthy, given that the Great Dividing Range is the only major mountain belt on the **continent**, separating coastal and inland **drainages** along the length of the eastern Australian mainland.[12]
- The Jagara hairy crayfish is **restricted** to the higher **altitude** rainforest community of Main Range National Park.
- *Euastacus* crayfish are distinguished from the smooth-shelled *Cherax* **species** by the short, **robust** spikes on their claws and **carapace**, and frequently, their larger size. Many *Euastacus* **species** grow to a relatively large size, with the Murray River crayfish (*Euastacus armatus*) being the second-largest freshwater crayfish **species** in the world. (The largest freshwater crayfish in the world is the Tasmanian giant freshwater crayfish (*Astacopsis gouldi*). Climate change, drought, bushfires, and rising water temperatures threaten it.[13]
- The **genus** *Euastacus* is **restricted** to the south-east of the Australian mainland. There is a high degree of endemism in *Euastacus* **species** in coastal river systems, with many **species restricted** to a single river or **creek** catchment. *Euastacus* **species** occur in several upland reservoirs.
- *Euastacus* **species** are extremely slow-growing, long-lived (possibly 40+ years in some **species**), and late to reach sexual **maturity**. These biological **characteristics** make *Euastacus* **species** vulnerable to **environmental** disturbances and essentially unable to support catch-and-kill fisheries.[13]

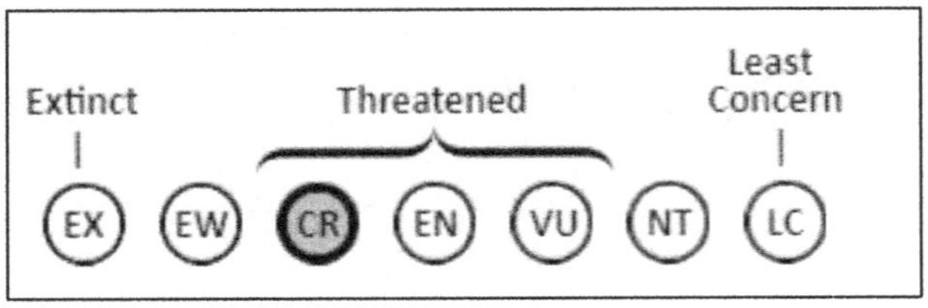

[1]

NCA List = CR = Endangered

EPBC List = CE = Critically Endangered

Large-eared pied bat

Chalinolobus dwyeri

- Large-eared pied bats are found in well-timbered areas with gullies. They are likely to **hibernate** during the coldest months. They are slow fliers and have been frequently spotted along **creek** beds and in the mid-**canopy** level.
- These bats have **glossy** jet black fur all over with a white band down the sides where the wing **membranes** meet and form a V-shape where they join in the pubic region. The ears are large, and they meet easily when pressed flat across the crown of the head. They have short, broad wings that make them manoeuvrable fliers.
- They usually **roost** in caves and other **crevices** in groups of three to 40 individuals. They remain loyal to the same cave over many years.
- They probably **forage** for small, flying insects below the forest **canopy**.
- Females commonly give birth to twins in November, where females gather in groups of 20-40 individuals in sandstone caves.
- Threats include **feral** cats, hawks and owls, land clearing, and wildfire. Disturbance and damage to **primary** nursery sites by animals and humans are also a confirmed threat. Alteration of **habitat** following subsidence due to longwall mining has been listed as a key threatening process in NSW. Much of the Large-eared Pied Bat's habitat occurs in sandstone underlain by coal seams, potentially at risk of collapse from underground mining.

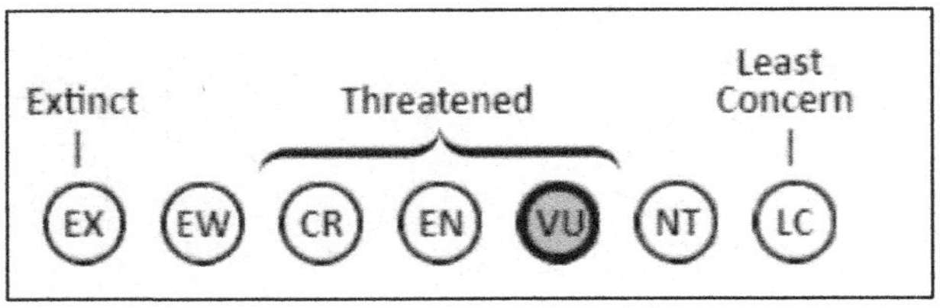

[1]

NCA List = V = Vulnerable

EPBC List = E = Endangered

Semon's leaf-nosed bat

Hipposideros semoni

- Also known as the greater wart-nosed horseshoe-bat, the noseleaf of Semon's leaf-nosed bat is well developed, square-shaped and covers most of the muzzle. There are two wart-like **protuberances** - one in the centre and another on the **posterior** edge. The upper portion of the noseleaf is divided into four depressions, and there are two supplementary leaflets under each side of the lower portion. The ears are particularly long and narrow, with an acute point. The forearm length is between 42 and. The fur is relatively long and has a ruffled appearance; it is dark, smoky-grey in colour, though lighter on the belly. The wing **membrane** near the body is covered with whitish-brown hair.[14]
- Semon's leaf-nosed bat is found in tropical rainforests, monsoon forests, wet sclerophyll forests, and open savannah woodlands. They appear to prefer rainforest and are more likely to be tree-dwelling than cave-dwelling
- There are three main population **isolates** in Australia identified for this **species**: Cape York, Kroombit Tops and Maryborough.
- Semon's Leaf-nosed Bat is considered to be in **decline** in Australia.
- Like many **species** of bat, this **species** takes advantage of disused underground mines, but is less reliant on mines than other **species**, and does not form large aggregations. Consequently, Semon's leaf-nosed bat has not been identified as a **species** that has experienced **significant range** expansion due to the creation of **habitat** from mining. They **roost** in caves, abandoned mines, rock fissures, and under rock overhangs. They prefer dark places.
- Heavily **pregnant** Semon's leaf-nosed bat females have been captured in October, and birth probably occurs in early November. A single pup is produced, though females have been observed with two stretched pubic teats, which might indicate twins.
- They are **nocturnal** and **roost** alone or in small **colonies**.
- Semon's leaf-nosed bats are **insectivorous** and use **echolocation** to locate **prey**. They **forage** in **dense** vegetation, hawking **prey** in the undergrowth within 1-2m of the ground, and gleaning **prey** from surfaces. They also fly low along forest gaps. They eat moths, beetles, and spiders. They hunt in small groups.
- They may be **preyed** upon by ghost bats, which are known to **consume** other bat **species**.
- They are **vulnerable** to human disturbance, mining, and **habitat** loss.

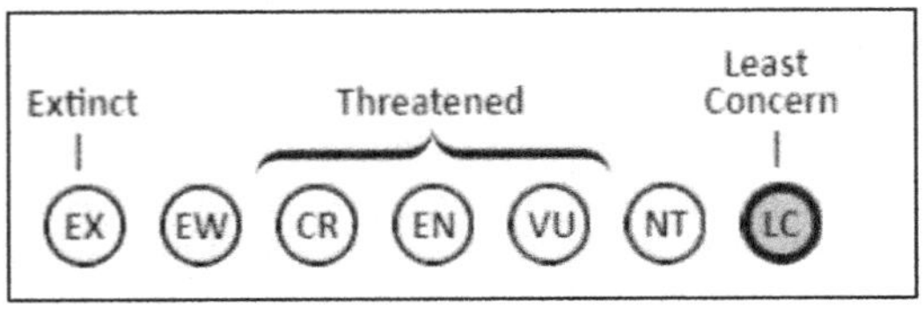

[1]

NCA List = E = Endangered

EPBC List = E = Endangered

Microbats

General information about microbats

- Bats are the world's second biggest group of **mammals**, and scientists are still finding new **species** of little microbats today. Approximately 70% of bats are microbats.
- Microbats are small, and the average body size is about that of a mouse. They are often called 'flying mice'.[3]
- Microbats weigh from 2 grams to 200 grams - a teaspoon of sugar weighs about 5-6 grams![3]
- All bats are **mammals**. **Mammals** are animals that care for their young after birth by feeding them milk. Most **mammals** give birth to live young. They all have fur or hair on their bodies. Bats are the only **mammals** that can fly.
- Bats' wings are long hands – like us, they have one thumb and four fingers.
- Bats are **nocturnal**. They wake up at **dusk** and fly out to feed and find water. They return to the **roost** early in the morning. Bats sleep or rest by hanging upside down, using a special claw on their foot to ensure they won't let go when they are asleep. The weight of the sleeping bat also causes the tendons in their legs to tighten. By hanging upside down, they don't waste any energy. In general, bats sleep upside down so they can easily spread their wings before taking off.
- Microbats can reach speeds of up to 160km per hour!
- A microbat's diet largely consists of insects but can include small **mammals**, frogs, and fish. Microbats can eat up to half their body weight in insects each night, equivalent to hundreds of insects per hour.
- Hearing is the most developed of all the senses - they can hear sounds up to 200,000Hz (we can only hear frequencies up to 20,000Hz)!
- Microbats do not rely on sight to find food. Instead, they use a form of radar called **echolocation**. Microbats **echolocate** by bouncing sound waves off objects and listening to the echo. They also have a magnetic sense – they can perceive the Earth's magnetic field and use it for orientation when flying.
- Most Australian microbats **roost** in tree hollows or under bark, or can be found in caves and cave-like structures. When there aren't enough 'natural' **roosts**, they can also find shelter in roofs or sheds.
- Microbats also have nurseries where the young are left in large groups while the mothers feed at night. In some areas, these nurseries can consist of millions of bats.
- Food can be **scarce** in winter, so microbats survive by either **migrating** to warmer regions or hibernating. **Hibernation** is a state of **torpor** (inactivity). Microbats can slow their heart rate and lower their body temperature, sometimes to as low as 2°C.

- The oldest known bat **fossil** is 52 million years old. Research shows it could fly but not use **echolocation**. Bats have been used as food sources, are a feature in Dreaming stories, are found in cave paintings and modern art, and are **totem** animals for some **indigenous** groups.
- Loss of **habitat** is the single most **significant** threat to all bats. **Predators** include cats, birds of **prey**, snakes, and, sometimes, it is thought, even other bats such as the broad-nosed bat (see page **10**).
- Like other wild animals, some bats may carry viruses that can be **transmitted** to humans. The two major viruses that have been identified in some bats are the Australian Bat Lyssavirus and the Hendra Virus - both can be **fatal**!
- If you find a sick or injured bat, **do not touch it**. Contact a local vet for details or your nearest wildlife carer.
- Microbat **habitats** include:
    - Shelter, such as trees or caves
    - Cave-dwelling microbats don't always need huge caverns to **roost** in. Even small rocky outcrops and **crevices** will do. However, **maternity colonies** need more room and can be found in larger caves.
    - Food – **abundant** insects to eat.
    - Water – microbats will wet their belly and lick their fur to get extra water that they don't get from their food
- Man-made **habitats**:
    - Houses
    - Farmland
    - Buildings in cities
    - Abandoned mines
    - Holes in old concrete drains can provide **habitat** for microbats.
    - Cave-dwelling microbats will use tunnels and abandoned mine shafts to **roost** in.
    - Microbats can often be found in houses or other buildings where they are safe and warm, and no other natural **roosts** are available.
- How can we help?
- Microbats need:
    - Forests, old trees, hollows.
    - **Abundant** food.
    - Places they won't be disturbed.
  - We can help by:
    - Stop clearing forests and old trees.
    - Installing bat boxes where old trees have been removed.
    - Reduce the use of **pesticides** around the home and on farms.
    - Know where microbats are **roosting** and leave them alone.
    - Learning more about microbats.
    - Telling others about microbats.

Bridled nail-tail wallaby

Onychogalea frenata

- The bridled nail-tail wallaby was declared extinct in 1937 and rediscovered some 40 years later, with only a few wild **populations** existing currently. As of 2001, the **range** of this wallaby was **estimated** at less than 5% of its original **range**.[15]
- The bridled nail-tail wallaby is found in three **isolated** areas in Queensland, with the only naturally occurring population confined to Taunton National Park, Queensland, with small, re**introduced populations** present at Idalia National Park and Avocet Nature Refuge in central Queensland. In early 2019, the total population of the **species** was **estimated** to be fewer than 500 **mature** individuals in the wild and 2285 in **captivity**.[15]
- These wallabies are named for two distinguishing **characteristics**: a white "bridle" line that runs down from the back of the neck around the shoulders, and a horny spur on the end of the tail. Other key physical features include a black stripe running down the **dorsum** of the neck between the scapulae, large eyes, and white stripes on the cheeks. Its fur is soft and greyish, with a darker tail tip. The bridled nail-tail wallaby can grow to one metre in length, half of which is tail, and weighs 4–8kg.[15] Females are somewhat smaller than males. The tail spur can be 3–6mm long and partly covered in hair. Its purpose is unclear.
- The **species** are most active during the nighttime and **dusk** periods. The day is usually spent sleeping in hollows near bushes or trees. In modern **habitats**, nail-tails keep close to the edges of pasture grasses. This wallaby prefers the transitional area between open grassy woodland and **dense** Acacia scrub. It's considered an edge specialist.
- This wallaby is mostly **nocturnal** but may be seen basking in the sun during cooler months. It's shy and **solitary** but may occasionally form small groups of up to four to feed when **grazing** is sparse.
- This wallaby eats a variety of grasses, shrubs, and herbaceous **species**.
- This wallaby can have up to three young per year. The **gestation** period is about 24 days, and the young stay in the pouch for about 120 days.
- This wallaby is threatened by predation from foxes, cats, wild dogs, and dingoes. It's also threatened by **habitat** loss, drought, and competition with **introduced** livestock. **Conservation** efforts include **predator** control, **habitat** protection, and fire management.
- A new strategy known as head-starting has been trialled at Avocet, with a successful outcome reported in May 2021: the population had grown by more than 100% over three years. A 10-hectare (25-acre) area had been fenced off three years earlier, with the youngest wallabies placed in it and allowed to grow in an area safe from **feral** cats, after studies had shown that it was overwhelmingly this group that was most vulnerable to predation. Although this method had been tested on **reptiles** and birds before, this was its first trial with land-based **mammals**. At Taunton Conservation Park, a total of 65 of the wallabies were recorded in 2008, but by early 2021, the population had risen to an **estimated** 1,300.[15]

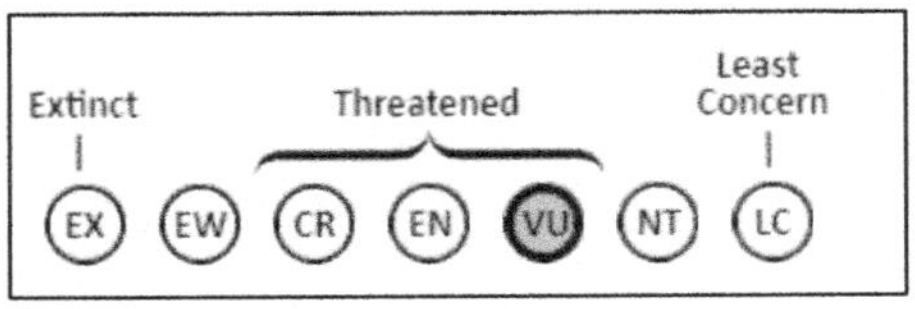

NCA List = E = Endangered
EPBC List = E = Endangered

Silver-headed antechinus

Antechinus argentus

- Antechinus are sometimes called "broad-footed **marsupial** mice", "pouched mice", or "Antechinus shrews".
- The silver-headed antechinus is one of the world's smallest **marsupials**, measuring 9-12cm in length (that is only slightly bigger than a business card). They are also one of the rarest **mammals** in Australia. They rank among the top 20 Australian **mammals** most likely to go extinct.
- It was first confirmed as a new **species** in April 2013.
- It is usually found in the wet eucalypt and rainforests of three national parks in Central Queensland. These are Blackdown Tableland National Park, Kroombit Tops National Park and Bulburin National Park.
- At Kroombit Tops National Park, the silver-headed antechinus occurs on an undulating sandstone plateau bounded on the eastern side by an escarpment with cliffs, at an elevation of 850-900m above sea level. At Blackdown Tableland National Park and Bulburin National Park, the silver-headed antechinus occurs in similar wet, high-**altitude** open forest **habitat**.
- Sadly, the devastating bushfires of 2019-2020 wiped out about 80% of its **habitat** in the **species**' stronghold in Bulburin National Park.
- Despite being so small, they have an impressive set of teeth which they use to eat spiders, cockroaches, beetles, and other bugs that they find whilst scurrying around in the thick vegetation they call home.[16]
- Males die out by the end of the mating season. This is because their testosterone levels are so high that when mating, their bodies pump their system full of the stress hormone cortisol, causing internal bleeding, hair loss and blindness. Despite all of this, the males continue to look for females to mate with until they inevitably die.
- In part due to its very small size, the silver-headed antechinus is challenging to find, so detection dogs, sensor cameras, and Elliot traps (small metal box traps) have been used to find the silver-headed antechinus.
- **Conservation** efforts: Protecting rainforest **habitats** from fire and **invasive** weeds - with **predictions** of drying and more intense fires due to climate change, we will have to work hard to save this tiny **carnivorous marsupial**. Controlling **feral predators** and other **feral** animals, such as pigs, **monitoring populations**, and researching their **habitat** preferences and distribution patterns will also be significant contributors to the **survival** of this **species**.

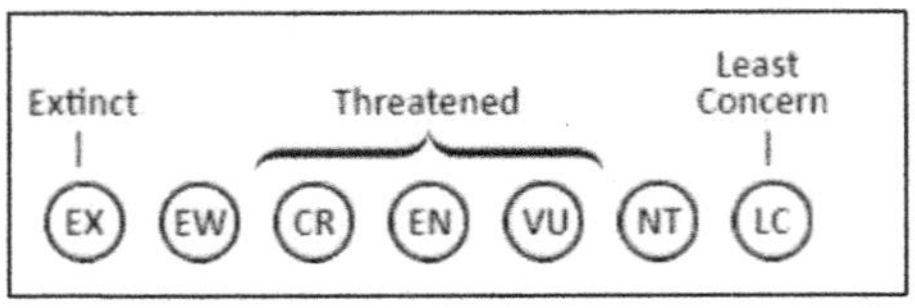

[1]

NCA List = E = Endangered
EPBC List = E = Endangered

Southern greater glider

Petauroides volans volans

- The southern greater glider is the largest gliding possum in eastern Australia. Its average body length is 35-45cm, and its long and furry tail measures 45-60cm. It weighs between 1 and 2kg.
- The southern greater glider could easily be mistaken for a fluffy tree-dwelling teddy bear, with its large, round ears and huggable, soft body. The thick fur increases its apparent size. The fur colour is white or cream below and varies from dark grey, **dusky** brown to light mottled grey and cream above.
- It has large ears with hair that projects past the outer edge of the ear.
- The southern greater glider occurs in eastern Australia, in eucalypt forests and woodlands, where it has a broad **distribution** from around Proserpine in Queensland, south through NSW and the Australian Capital Territory into Victoria
- The southern greater glider feeds exclusively on eucalypt leaves, buds, flowers and mistletoe.
- It shelters during the day in tree hollows and will use up to 18 hollows in its home **range**. They are recorded using hollows with a minimum diameter of 8cm. These hollows can help regulate its body temperature and keep it safe from **predators**.
- They occupy a relatively small home **range** with an average size of 1-3ha.
- Southern greater gliders give birth to a single young in late autumn or early winter, which remains in the pouch for approximately 4 months and is independent at 9 months of age.
- They are usually **solitary**, though mated pairs and offspring will share a den during the breeding season and until the young are independent.
- Very loyal to their **territory**.
- Southern greater gliders can glide up to a **horizontal** distance of 100m, including direction changes of as much as 90 degrees.
- Threats include:
    - Loss of **habitat**, **habitat fragmentation** due to human activities including farming, forestry, **urbanisation** and rural development.
    - Loss of hollow-bearing trees (which can take 300-400 years to develop hollows).
    - Alteration of microclimate conditions due to climate change, changes in rainfall, prolonged droughts, increased temperatures and heat waves. This **species** requires sufficient **moisture** on tree leaves to metabolise the toxin content.
    - Frequent or high severity fires.
    - Small population size making them **susceptible** to loss of individuals from predation and threats impacting **habitat** quality and food availability.
    - Barbed wire fences can entangle greater gliders and damage their gliding **membranes**.

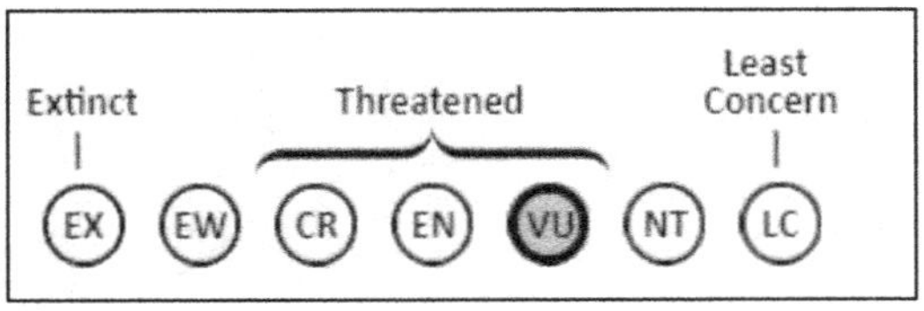

[1]

NCA List = V = Vulnerable
EPBC List = E = Endangered

Spotted-tailed quoll (southern subspecies)

Dasyurus maculatus maculatus

- Early settlers referred to quolls as '**native** polecats', because they resembled the **European** relative of the weasel, and today they are still sometimes called **native** cats.
- Quolls are members of the *Dasyuridae* family, and are easily **recognised** by their pointed **snout**, long tail and ginger/brown or black fur with distinctive white spots. They have sharp teeth and strong jaws for tearing meat, and crushing bones and **invertebrates**.
- Like the related Tasmanian devil, these fierce little **marsupials** communicate loudly with hisses and screams.
- The spotted-tailed quoll is the largest of the quoll **species,** about the size of a domestic cat, with males weighing up to 5kg and females up to 4kg.
- The spotted-tailed or tiger quoll is mainland Australia's largest **carnivorous marsupial**.
- The spotted-tailed quoll has reddish-brown fur, white spots on its back and tail, and a pale belly.
- The spotted-tailed quoll is found in eastern Australia, including Queensland, Victoria, and in Tasmania.
- The spotted-tailed quoll faces many threats, including **habitat** loss, competition with other animals, deliberate **poisoning**, shooting, and trapping.
- The spotted-tailed quoll population has **declined** since **European settlement**.
- The spotted-tailed quoll is highly mobile, travelling up to several kilometres per night.
- Quolls shelter in hollow logs or dens during the day, mainly emerging at night to hunt. The largest **species**, the tiger quoll, eats birds, **reptiles** and **mammals** such as bandicoots, possums and echidnas, while the smaller **species** eat mainly insects, birds, frogs, lizards, snakes, small **mammals** and fruit. Quolls are also known to eat **carrion** and will scavenge around campsites and rubbish bins.
- Quolls are generally **solitary**, except during the breeding season and when they frequent unusual communal 'latrines' to mark territories and **socialise**.
- The spotted-tailed quoll is **preyed** upon by **native** and non-**native predators**, including red foxes and **feral** cats.
- The spotted-tailed quoll can be **poisoned** by eating cane toads in the wild or by ingesting 1080 **poisoning** baits.

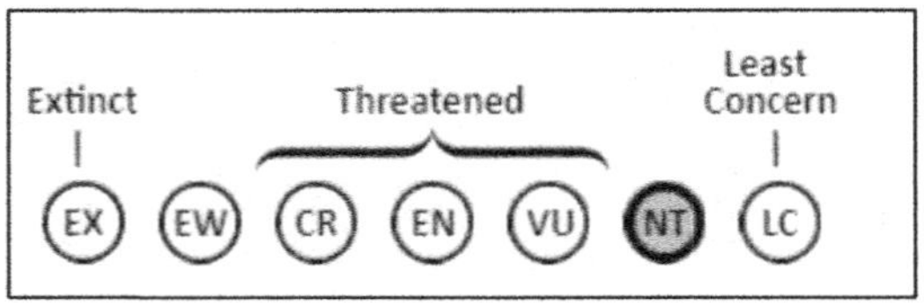

[1]

NCA List = V = Vulnerable
EPBC List = E = Endangered

Gouldian finch

Chloebia gouldiae

- Gouldian finches are known for their striking colours, which include green backs, yellow bellies, and purple chests. They also have black, red, or yellow faces. Gouldian finches have colour morphs linked to behavioural **characteristics** and mate selection.
- Once found across northern Australia in their millions, in recent decades the population of the Gouldian finch has **declined** dramatically to an **estimated** 2,500 **mature** individuals. The **species** now only occurs in small **flocks** in a few locations in the Kimberley and the Northern Territory, although it is occasionally recorded in Queensland.
- Gouldian Finches live in **tropical** savanna woodland amongst the grassy **understorey** that contains a mixture of their preferred **annual** and **perennial** grass **species,** usually near water.
- Gouldian finches are herbivores that eat almost entirely grass seeds and nest in tree hollows.
- Gouldian finches build loosely woven grass nests in tree hollows or termite mounds.
- Gouldian finches are very **social**, forming **flocks** of hundreds during the non-breeding season.
- Gouldian finches are **preyed** upon by snakes, **feral** cats, and hawks.
- The **primary** threat to the **survival** of the Gouldian Finch is altered fire regimens, especially an increase in **extensive**, hot wildfires in the late dry season. Wildfires reduce seed availability at key times during the year and limit the availability of tree hollows for nesting.[17] Cattle **grazing**, **invasive species**, wildfires, and lung **parasites** all threaten the Gouldian finch.
- Gouldian finches are popular pets, but captive **populations** differ from wild ones due to selective breeding.
- Gouldian Finch chicks have luminous nodules called papillae on the sides of their beaks and markings on the inside of their mouths. They act as a sort of "beacon" to direct the parents to the hungry mouths of their chicks in dark nest cavities.

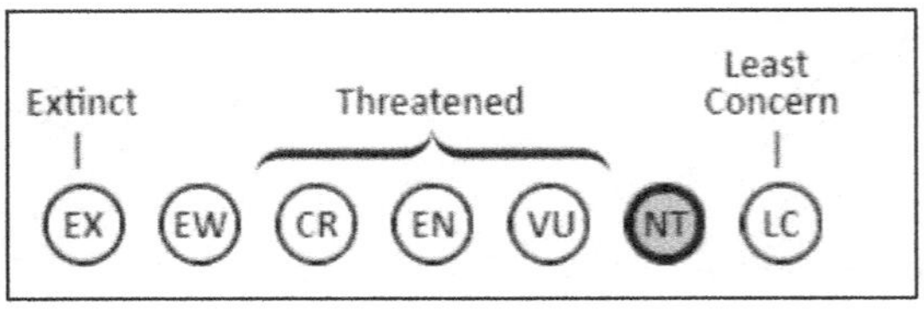

[1]

NCA List =E = Endangered
EPBC List = E = Endangered

Major Mitchell's cockatoo

Cacatua leadbeateri leadbeateri

- With its soft-textured white and salmon-pink **plumage** and large crest with bands of scarlet and yellow, it is often described as the most beautiful of all cockatoos. Standing at around 36cm, it is one of the smaller cockatoos.
- It is named in honour of Major Sir Thomas Mitchell, an explorer and surveyor of southeast Australia in the 1800s.
- Males and females are almost identical. The males are usually bigger. The female has a broader yellow stripe on the crest and develops a red eye when **mature**.[18]
- The bird reaches sexual **maturity** around 3–4 years old. The oldest recorded Major Mitchell's cockatoo died at 83 years old.[19]
- They are **omnivores** that eat seeds, nuts, fruits, roots, and insects. They are especially fond of acacia and cypress pine seeds.
- They live in wooded areas and around waterholes. In contrast to those of the galah, **populations** of Major Mitchell's cockatoos have **declined** rather than increased as a result of man-made changes to the arid interior of Australia. Where galahs readily occupy cleared and part-cleared land, Major Mitchell's cockatoos require **extensive** woodlands, particularly favouring conifers (*Callitris* spp.), **sheoak** (*Allocasuarina* spp.) and eucalypts. Unlike other cockatoos, Major Mitchell pairs will not nest close to one another, so they cannot **tolerate fragmented**, partly cleared **habitats**, and their **range** is contracting.[19]
- In the Mallee region of Victoria, where the galah and Major Mitchell's cockatoo can be found nesting in the same area, the two **species** have occasionally interbred and produced hybridised offspring.
- They are noisy birds that scream and cry. They are also very needy and require a lot of attention and stimulation.
- They mate for life and return to the same breeding grounds each year. They typically produce 2–4 chicks in the spring.
- They are threatened by **habitat** loss due to clearing for **agriculture** and **illegal** capture for sale. Wildfires also destroy their **habitat** and food sources.

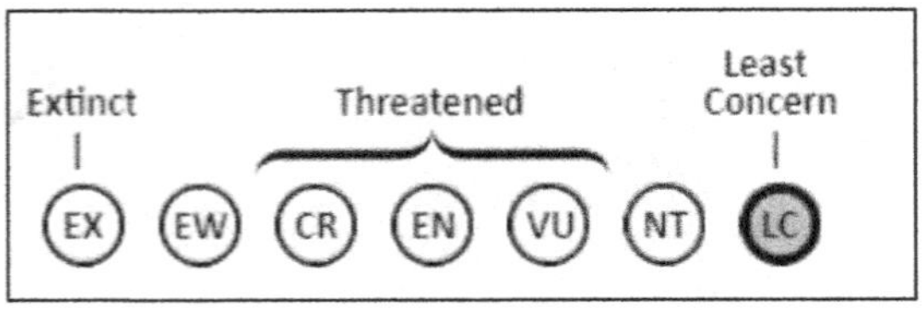

[1]

NCA QLD List = E = Endangered
NCA NSW = V = Vulnerable
NCA VIC = CE = Critically endangered
EPBC List = E = Endangered

Eungella mint bush

Prostanthera eungella

- The Eungella mint bush is a species of flowering plant in the family *Lamiaceae* and is **endemic** to the Eungella region in Queensland. It is an erect shrub with narrow egg-shaped leaves with small teeth, and mauve flowers that are white inside the petal tube and arranged in upper leaf axils.
- This mint bush is only known from near the type location where it grows in open forest and on the edge of rainforest.
- A small to medium **dense** shrub with highly aromatic foliage continuously releasing a mint essence into the atmosphere.
- A lavish display of violet flowers all but hides the soft leaves throughout spring and early summer, attracting butterflies and small insect-eating birds.

NCA List = CR = Critically endangered
EPBC List = C = Critically endangered

No common name is known

Acacia deuteroneura

- *Acacia deuteroneura* is a shrub to 3m high with hairless, waxy, white branchlets. Phyllodes are blue-green, 3–6cm long, 6–12mm wide, blunt at the apex, leathery, hairless, with a **prominent** midrib and a minor nerve running parallel to the midrib for two thirds of the length of the phyllode. Flowers are yellow, 20–30 in nearly spherical heads. The **inflorescence** axes and stalks bearing the flower heads are hairless. Pods are to around 5cm long, 12–13mm wide.

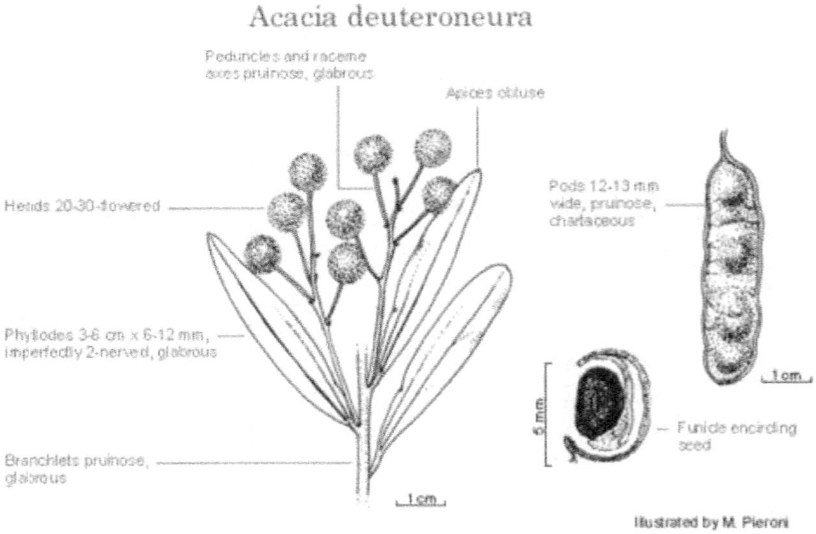

- Seeds are 3.5–4mm long, black, borne on a stalk that is folded and encircles three quarters of the seed.
- It is known from two locations: one 64 km north-north-east of Tambo in central western Queensland, growing on a knoll of weathered sandstone with *Eucalyptus bakeri* and the other from 60 km north of Tambo. There are only two **specimens** of *A. deuteroneura* in the Queensland Herbarium, collected in 1968. The collecting notes for the 1973 **specimen** describe the species as "forming almost pure stands".

NCA List = CR = Critically endangered
EPBC List = V = Vulnerable

Lloyd's native olive

Notelaea lloydii

- Lloyd's **native** olive is a shrub which typically grows to a height of 1-3m with leathery, linear or slightly sickle-shaped leaves, pale yellow or cream-colored flowers with 4-lobed petals, 2 **stamens** and a **glabrous ovary**. The fruit is a dark blue **drupe**.
- It is known from eight sites at five locations within southeast Queensland.
- The species occurs on undulating to hilly terrain either in **moist** gullies or on gentle to steep dry stones, but it is rarely found on rocky outcrops.
- The main **threat** to Lloyd's Olive is habitat loss and **degradation** due to urban development. The current fragmented, **rare** and **isolated** nature of the species means it is more vulnerable to loss of remaining habitat. Additionally, inappropriate fire regimes are a potential **threat** to the species.

NCA List = V = Vulnerable
EPBC List = V = Vulnerable

Mount Stuart Mystery Tree

Backhousia tetraptera

- Tree, often multi-trunked, usually to 5m but occasionally to 12m, coppicing.
- It is known only from an **isolated** population near the summit of Mt. Stuart, Townsville.
- Leaves are simple, lance shaped to oval, bark, flaky thin grey to mottled grey-brown to pink, flowers, white, terminal or in upper axils in clusters of 10 with numerous **stamens**. Fruit, dry, indehiscent 4-winged capsules, pink fading to white and brown in compact clusters.
- Grows in deciduous vine thickets and in gullies protected from fire.
- The current fragmented, **rare** and **isolated** nature of the species means it is more vulnerable to loss of remaining **habitat**. Myrtle rust poses a serious **threat**. Additionally, inappropriate fire regimes are a potential **threat** to the species. Drought can further exacerbate the impact of other **threats**.

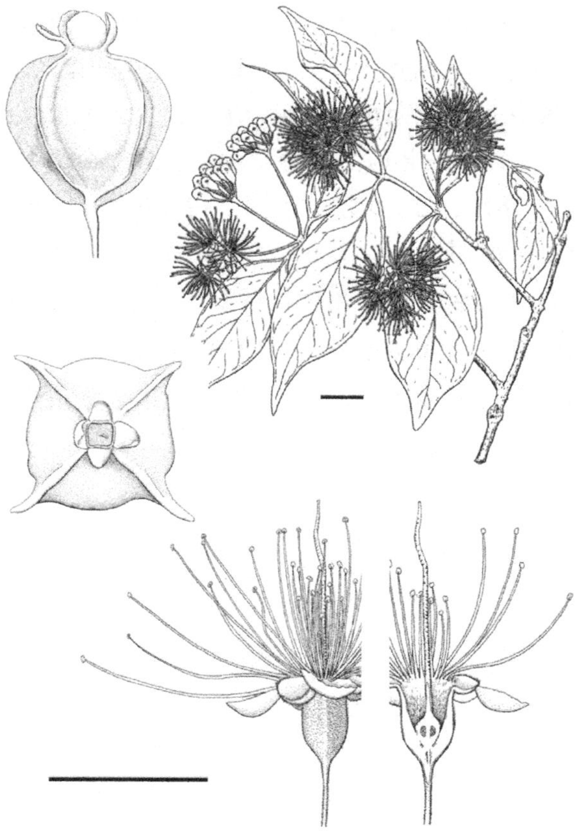

NCA List = CR = Critically endangered
EPBC List = CE = Critically endangered

White burr daisy

Calotis glabrescens

- This small white daisy grows 15-20cm tall and is listed as critically endangered under the Queensland NCA but is not currently listed under the EPBC Act; it is being **monitored** and assessed.
- Until recently, this **species** was listed under the NC Act as presumed **extinct**, as there were only two records. One **specimen** had been collected in 1946 from near Bybera, in what is now Whetstone State Forest, noting it was 'moderately common in open forest land'. The next collection was in November 1997 by the Queensland Herbarium, who noted it was a 'low herb to 10 cm tall with pinkish **inflorescences**, uncommon, scattered at site'. This collection came from approximately. 18km south-west of Leyburn. The **species** was not collected again after 1997 until two records in 2020 from Durikai State Forest and another site south-west of Leyburn.
- Collection records for the **species** associate it with open forest and semi-cleared areas containing **species** such as *Eucalyptus moluccana, E. sideroxylon, E. viridis* and the wattles *Acacia montana* and *A. ixiophylla.*
- The 1997 collection notes also associate it with 'traprock' – a type of metamorphic rock known from the Warwick area.
- In October 2021, a survey found 15 plants growing on an undulating plain beside a dirt track, on a hard-setting pale clay loam, on the edge of a large clearing (but in the clearing, not in the adjacent woodland). Adjacent woodland was co-dominated by *E. moluccana* and *E. dealbata,* and was relatively low (median height of 10-15m). The parent rock was metamorphic, and the pre-clear regional **ecosystem** was *Eucalyptus melliodora* and/or *Eucalyptus microcarpa/ E. moluccana* woodland on rolling hills, depressions and lower slopes around **drainage** lines.
- This **species** is difficult to survey for because it is physically small and may be dwarfed by the grasses and other herbs with which it occurs. In addition, it may be present with other **superficially** similar *Calotis* spp., which may be **abundant**.

NCA QLD List = CR = Critically endangered
EPBC List = Not currently listed

Koala

Phascolarctos cinereus

NATIVE TO AUSTRALIA

- The word "Koala" is thought to originate from one of the Australian **Aboriginal** languages and means 'no drink'. Koalas drinking water is uncommon although they may do during heat waves, droughts or bush fires.
- The average koala weighs about 9-10kg. They have five **digits** on each paw, but on the front paws, two are **opposing**, like thumbs, to help them grip trees and branches better.
- Koalas and humans have almost identical fingerprints!
- They have a specialist diet of gum leaves, **poisonous** to most other animals. They can eat up to 1kg of leaves a day. Koalas have an excellent sense of smell and will sometimes sniff leaves before tasting them, favouring the more **nutrient-rich** leaves.
- They can sleep for 18-20 hours a day. **Digesting** all those leaves takes a long time, and they do not get a lot of energy from just eating leaves.
- They are the only **surviving** members of their family; their closest living relatives are wombats.
- They don't live in **social** groups.
- Koalas are **marsupials**.
- **Did you know?** Female koalas are only **pregnant** for around 35 days. In comparison, an elephant is **pregnant** for approximately 620 days![20]
- Baby koalas are called **joeys**. They are born without hair, blind and without ears, and are about the size of a jellybean.
- When ready to leave the pouch, a **joey** will hitch on mum's back for six months to learn **critical survival** skills such as **foraging** for food and avoiding **predators**.[20]
- Koalas are usually slow-moving but can gallop up to 30km/hr when threatened.
- The "Black Summer" bushfires in 2019-2020 are thought to have been responsible for the deaths of 80% of koala **populations** in some regions. During bushfires, koalas unable to escape burning areas can be killed directly by flames, their **habitats** are destroyed, food availability is reduced, and there is an increased risk of predation. Such **traumatic** events also impact stress levels and **disease susceptibility**, disrupt breeding patterns, and cause **fragmentation** of **populations**, **hindering genetic diversity**.

Watch this short video from ABC iview on YouTube:
The beautiful bond between koala and eucalyptus tree | Nature's Great Divide
https://www.youtube.com/watch?v=QC2V311zieo

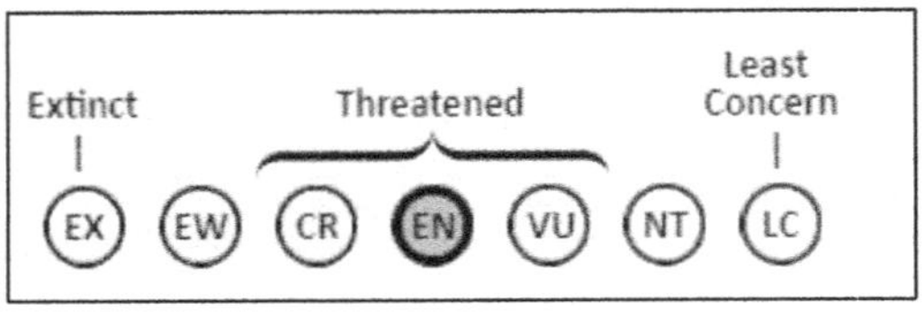

[1]

NCA List = E = Endangered

EPBC List = E = Endangered

Black-throated finch (white-rumped subspecies)

Poephila cincta cincta

- The black-throated finch is found in grassy woodlands throughout northeastern Australia from Cape York Peninsula to central Queensland. The southern black-throated finch (white-rumped **subspecies**), found south of Townsville (*Poephila cincta cincta*), is **endangered**, with a population in **decline** and its **habitat** is threatened by development, and has become **extinct** in New South Wales. In contrast, the northern black-throated finch (black-rumped **subspecies**), found north of Cairns (Poephila cincta atropygialis), is not listed as threatened at this point.
- The Latin *cincta* may be interpreted to mean "girdled".
- Measuring around 10cm, the black-throated finch has a short black **bill**, lores, and throat, sharply delineated from the rest of the pale grey head. The wings, breast and belly are pale pinkish brown, the short tail is black, while the **rump** is black in northern forms and white in southern forms.[21]
- The vocalisation of *P. cincta* is comparable to that of the sister **species** *Poephila acuticauda*, although lower in tone and slightly less simple in the harmonic structures. Up to twelve calls have been identified, and the structure and tone of these are also distinguishable by **subspecies**.[21]
- The black-throated finch was traditionally found from Cape York south through eastern Queensland and into north-eastern New South Wales in the vicinity of Tenterfield; however, it has not been recorded in New South Wales since 1994. It is **sedentary** or **locally nomadic**. It is found in grassy, open forested **habitats**, generally near bodies of water, such as rivers.
- The black-throated finch is found in **flocks** of up to 30 birds. The behaviour is similar to that of the longtail *Poephila acuticauda*, with close pair bonds that are **isolated** or associating in loose groups of six or more pairs. They **forage** in a **range** that centres on the nest site, which is used for breeding or **roosting** outside of the **maternity** season. **Local** and seasonal conditions may cause pairs to join **congregations** where there are limited resources, such as food and especially water, and they may join large **flocks** to travel to water sources during drought.[21]
- The black-throated finch primarily eats seed from various **species** of grass and also eats spiders and ants. They **forage** on the ground for fallen seeds of **native** plants, their **primary** diet, and select these ones after the other. As with some other estrildids, they have also been observed **harvesting** seed still attached to the plant. They may occasionally bend the stem down, perhaps clasping it against the ground with their foot, to reach and select seeds from the seed-heads. Larger **flocks** may form to feed at recently burnt sites. The **species** also hunt termites flying during the breeding season, capturing them as they trail along the ground or when in flight; the immature finches eventually learn to shake the wings off before consumption. Spiders are plucked from their webs for an occasional supplement to their staple diet.[21]
- They can suck water by partially opening and immersing their **bill**, drinking in the morning or evening unless the water is readily available.
- For the past few decades, the population of this **species** has **declined**. While the northern **subspecies** (black-rumped, *Poephila cincta atropygialis*) has a status of "least concern" under Queensland's NCA, the Commonwealth status of the southern **subspecies** (*Poephila cincta cincta*) under the EPBC is "**endangered**". The **population decline** is probably due to the spread of **pastoralism**, changes in fire regime, and increases in the density of **native** woody weeds in grassy **savannahs**.[21]

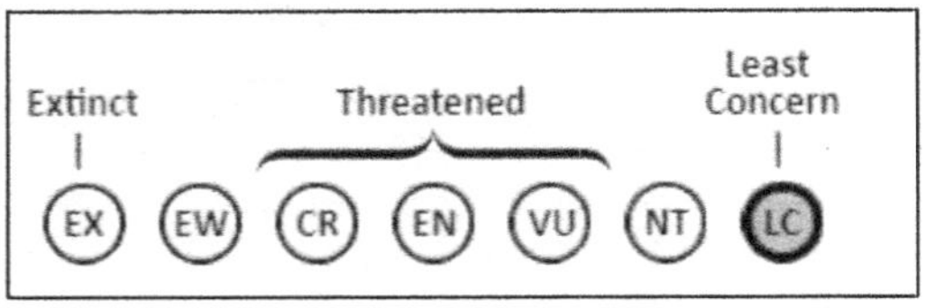

[1]

NCA List = E = Endangered
EPBC List = E = Endangered

Albert's lyrebird

Menura alberti

- Albert's lyrebirds are similar to superb lyrebirds, but smaller and with a richer brown **plumage**. Also known as the northern lyrebird, Albert's lyrebird is a timid, pheasant-sized songbird endemic to Australia's subtropical rainforests, in a small area on the state border between New South Wales and Queensland.
- Albert's lyrebird is named after Prince Albert, the prince consort of Queen Victoria, queen of the United Kingdom.
- They lack the elegant lyre-shaped tail feathers of the superb lyrebird and are found in a much more **restricted range**.
- They are ground-dwelling birds with females reaching approximately 75cm in length and males 90cm in length. The wingspan is 76-79cm and weigh about 930g. Male and female are alike except for the shape of the tail: the male has a spectacular tail composed of a central pair of long ribbon-like dark-brown median plumes, six pairs of long, filmy and luxuriant filamentary feathers, which are black-brown above and dark grey below; and a long broad fully webbed outermost pair of lyrates, which are black-brown above and dark grey below.[22]
- The total population of Albert's lyrebirds was **estimated** at only 3,500 breeding birds in 2000, with one of the smallest **distributional ranges** of any bird on the **continent**.[22]
- Both this **species** and the superb lyrebird have powerful, flexible voices. They are renowned for their impressive ability to **mimic** various sounds. The voice can create sounds at one moment deep and resonant, switch to high, thin squeaks and trills, then change again to harsh noises. They sing both their own songs and **mimic** the sounds of other birds, including alarm calls and the sounds of **predators**.[22]
- Male Albert's lyrebirds perform a **unique** dance during courtship, spreading their tail forward and shivering while calling loudly.
- They are considered "**ecosystem** engineers" as they rake the forest floor while **foraging**, potentially reducing bushfire fuel.
- The major threats to Albert's lyrebird include the intense management of forests and the replacement of optimal **habitat** with plantations of unsuitable **species**, such as eucalypts or hoop pines; invasion of logged or otherwise damaged **habitat** by weeds, especially *Lantana camara*, which reduces suitability of the **habitat**; damage to **habitat** by **grazing** stock; encroachment of urban or rural development close to **habitat** of Albert's lyrebirds; and predation by **introduced** red foxes (*Vulpes vulpes*), **feral** dogs and cats, and domestic dogs and cats, where the birds are located close to human **settlements**. Because the **range** of the **species** is confined to such a small geographic area, a **threatening** event, such as a severe regional drought, has the potential to affect all individuals.[22]

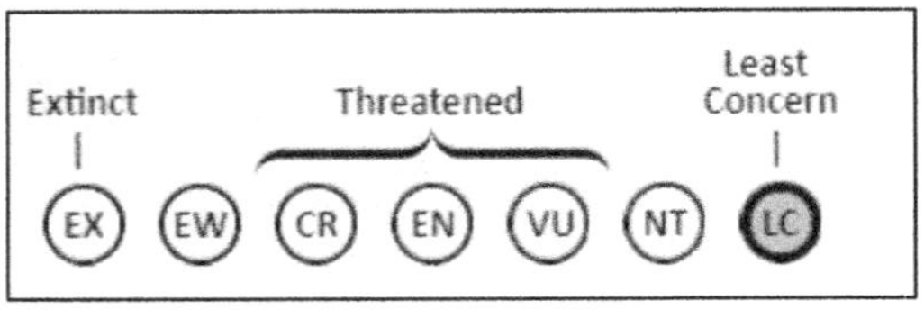

[1]

NCA QLD List = NT = Near threatened
BCA NSW List = Vulnerable
EPBC List = E = Endangered

Northern hairy-nosed wombat

Lasiorhinus krefftii

- The northern hairy-nosed wombat is the largest of Australia's three wombat **species** and is also one of the **rarest mammals** in the country. The northern hairy-nosed wombat is very **rare**, and the few **surviving** individuals live in Epping Forest National Park in central Queensland.
- Northern hairy-nosed wombats can live for up to 30 years in the wild, according to the Australian Wildlife Conservancy.
- They can grow over a metre long, weigh up to 40kg, and are covered in soft, greyish fur. They have long, pointed ears and a wider muzzle than the common wombat.
- The northern hairy-nosed wombat can move fast, up to 40 km/h over a short distance.
- The northern hairy-nosed wombat lives in drier grasslands and builds large tunnel systems called warrens in deep sand. They are **solitary** animals and also territorial.
- They are expert **burrowers**, spending much of their time underground. They are primarily **nocturnal**, spending long periods in their **burrows**, especially during the hot summer months.
- They use their powerful forelimbs and long claws to **excavate extensive burrow** systems, which can be up to 3.5 meters underground.
- Their pouches face backward to keep dirt out while digging, and they produce cube-shaped poo which is thought to help them mark their **territory** and prevent it from rolling away.
- The northern hairy-nosed wombat is an herbivore that eats grasses, roots, and bark.
- While not always **aggressive**, they can use their powerful hind legs and teeth to **defend** themselves if they feel **threatened**.
- The northern hairy-nosed wombat is a **marsupial**, which means its babies develop in its mother's pouch.
- Koalas are the closest living relatives of wombats, sharing similar physical **characteristics** and also having backwards-facing pouches.
- With the **species threatened** by wild dogs, the Queensland Government built a 2km **predator**-proof fence around all wombat **habitat** at Epping Forest National Park in 2002. Insurance **populations** have since been translocated to two other locations to ensure the **species survives threats** such as fire, flood, or **disease**. In 2003, the total population comprised 113 individuals, including only around 30 breeding females. After recording an **estimated** 230 individuals in 2015, the number was up to over 300 by 2021, and over 400 by 2024.

[1]

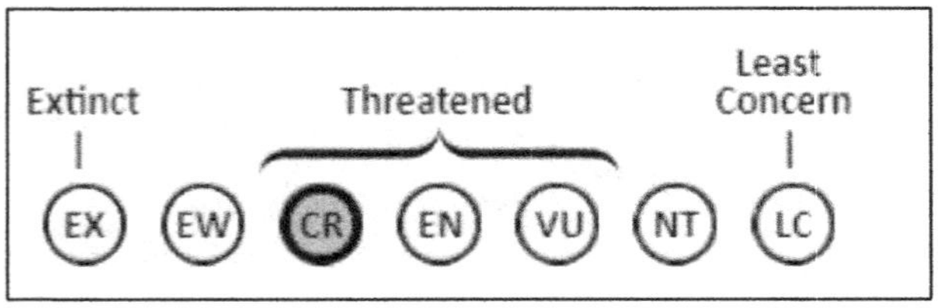

NCA QLD List = CR = Critically endangered
BCA NSW List = EX = Extinct
EPBC List = CE = Critically endangered

Merten's water monitor

Varanus mertensi

-
- This **species** of water monitor grows to a total length, including the tail, of about 1m. They are dark brown to black above with many cream to yellow spots. Their underparts are paler, white to yellowish, with grey mottling on the throat and blue-grey bars on the chest.
- Their tails are about one and a half times the length of their head and body, and are flattened and strong, helping them push through the water quickly so they can swim really fast.
- Mertens' water monitors spend much of their time near or in water, swimming, basking on rocks and logs, and branches overhanging **swamps**, lagoons and waterways. When disturbed, they will drop into the water where they can stay submerged. While they love the water, water monitors can also climb trees surprisingly well.
- < UNK> Like crocodiles, they have nostrils on top of their heads, so they can breathe while swimming and stay partially submerged when searching for **prey**.
- They have a special valve in their nostrils that shuts tightly when they go underwater, keeping the water out and allowing them to stay under the surface for several minutes.
- They are carnivores and eat fish, frogs, small **mammals**, insects, and even **carrion**.
- They have a good sense of smell and may dig up **prey** when **foraging**, including the eggs of freshwater turtles. They are **opportunistic predators** of **aquatic** and **riparian habitats**. They can swallow prey underwater, an ability not reported in any other monitor **species** other than the Borneo earless monitor (*Lanthanotus borneensis*).
- Mertens' water monitors have a long, forked tongue, similar to snakes, which they use to detect scents in the air, even underwater. This forked tongue acts as an important sensory organ, allowing them to "taste" the air and locate **prey** or other cues. The forked structure helps them sense the direction and intensity of scents, almost like "stereo hearing. "They are found in the far northern waters of coastal and inland Western Australia, the Northern Territory, and central and northwestern Queensland.
- They build **burrows** along riverbanks and may use termite mounds or stumps to lay their eggs, with egg-laying taking place early in the dry season and **hatching** in the following wet season. Depending on temperature, the eggs hatch within 200-300 days after laying, with the **hatchlings** able to enter the water and swim immediately.

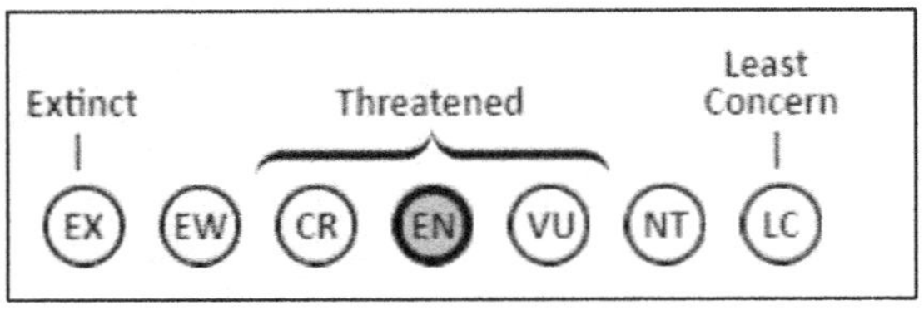

[1]

NCA QLD List = E = Endangered
EPBC List = E = Endangered

Boggomoss snail

Adclarkia dawsonensis

- The Boggomoss snail is an exceptionally **rare**, air-breathing land snail with only two known **populations** located within a 90 km stretch of the Dawson River valley between Taroom and Theodore in Queensland, Australia. One of these **populations** occurs in a boggomoss (aquifer-fed **moist environment**) on Mt Rose Station[23], and the other as a series of **subpopulations** in the **riparian environments** of the Dawson River.
- It is considered that the Brigalow communities, which once flourished on the alluvial black soils in this region, formed the historical core of the snail's **distribution**. Most of these communities have since been cleared for **agriculture** and cattle **grazing,** and the snail is now largely **restricted** to the **moister environments** of **riparian** zones and some boggomoss **habitats** that remain on these alluvial flats. Here it prefers the microhabitat of deep, damp layers of leaf litter, particularly around the base of trees such as eucalypts and sandpaper figs. These strict **habitat** requirements confine the snail's **distribution** to an area less than 10km^2.[23]
- They are known for their thin, **semi-transparent**, flattened shells, which can appear yellow to light brown. Black blemishes may be visible through the shell, reflecting markings on the lung roof.
- Shells of **mature** individuals are approximately 26mm wide and 16mm high.
- Interestingly, it **hibernates** during dry periods by sealing itself to the underside of logs.
- These snails have a preference for and thrive in **moist environments** with good **canopy** cover, fallen logs, and deep leaf litter.
- During dry spells, they seal themselves to the underside of logs or leaf litter to survive.
- They are **nocturnal** and mainly active at night.
- They are believed to feed on fungal and decaying vegetable matter.
- Their main **threats** include land clearing, fire, and other human-related impacts like cattle **grazing**. Other impacts include weeds, earthworks, and interruption to water flow.

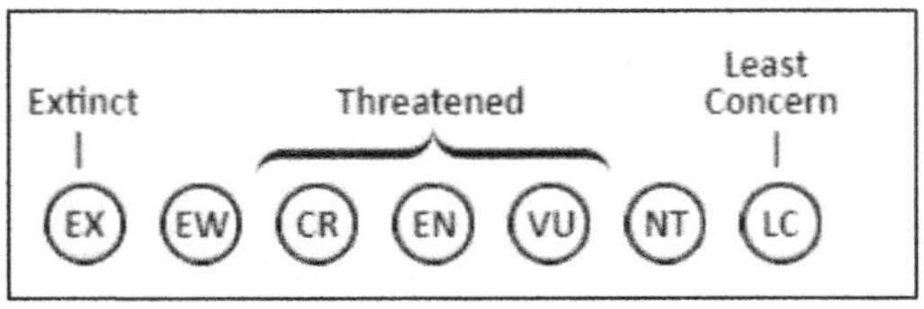

[1]

NCA List = E = Endangered
EPBC List = CE = Critically endangered

Masked owl (northern subspecies)

Tyto novaehollandiae

- The Australian masked owl is a fascinating bird of **prey** with several notable **characteristics**. It's the second-largest **nocturnal raptor** in Australia, known for its distinctive facial disk and powerful talons.
- There are nine **recognised subspecies** of the Australian masked owl.
- Masked owls are among the largest owls in Australia, with a wingspan of up to 129cm and a body length of 47-51cm. Females are generally darker and larger than males.
- A **prominent**, heart-shaped facial disk with a dark ring is a key feature of this owl. They have large, dark eyes. They have well-feathered legs, large claws, and strong feet.
- They often have a stocky, crouched posture.
- Masked owls are **nocturnal**, hunting primarily in the early night hours. Females often hunt **prey** that is a lot larger than what the males hunt. This makes sense given that the females can be much bigger than the males.
- They are often found in timbered areas with a shrub understory. They nest in large tree hollows near **foraging** areas.
- Their diet includes small **mammals** like **rodents**, rabbits, bandicoots and possums. They also hunt possums, **reptiles**, birds, and insects.
- The masked owl's diet comprises small **mammals**, **reptiles**, and birds.
- Like other raptors, these owls regurgitate pellets, which are soft clumps of **undigestible** bits of their **prey**, such as bones, claws, and fur. At one Australian masked-owl nest site, biologists found around 20kg of pellet material that had accumulated in and around the nest after many years of use![24]
- They are territorial and may remain in the same area once they have established a breeding **territory**.
- They usually lay two to three eggs. The female occupies the nest for up to 10 weeks before laying.
- They are **threatened** by loss of nesting **habitat**, clearing of **habitat** for development, and secondary **poisoning** from **rodenticides**. The masked owl population on the mainland is **declining**, with several states listing it as **threatened**.

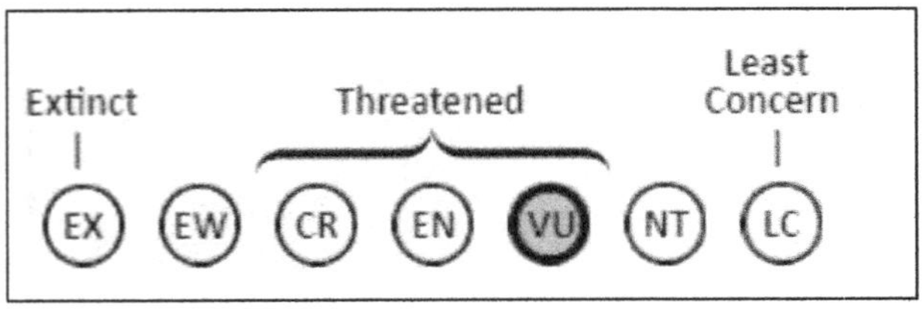

[1]

NCA List = LC* = Least Concern
*** At subspecies level this classification varies.**

EPBC List = V = Vulnerable

See page 118 (Powerful owl) for some interesting information about owls' eyes!

New Holland mouse

Pseudomys novaehollandiae

- New Holland mice, also known as Pookila, are small **native** Australian **rodents** with some **unique** features. They have a distinctive bicoloured tail (grey on top, white on the bottom) and pink feet.
- They have large eyes and rounded ears, and lack the "mousey" odour of house mice. Their eyes and ears are also larger than those of the house mouse.
- Females can live for up to 2 years in the wild.
- The New Holland mouse has a **fragmented distribution** across Tasmania, Victoria, New South Wales and Queensland.
- They are primarily found in **heathlands** and coastal areas, although they **inhabit** various **environments** including woodlands, forests with **heathland understories**, and vegetated sand dunes.
- New Holland mice are **omnivores**, with their diet consisting of **native** plant seeds, leaves, flowers, fruits, **fungi**, and some **invertebrates**.
- New Holland mice are **nocturnal**.
- They are communal nesters, living in underground **burrows** with other mice during the day.
- They are thought to play a role in seed and **fungal spore** dispersal, impacting plant **reproduction** and **ecosystem** health. They also contribute to healthy **ecosystem**s through soil turnover.
- New Holland mice are **declining** in numbers due to climate change, drought, **habitat** loss, and **predators**.
- The total population size of **mature** individuals is now **estimated** to be less than 10,000. However, given the number of sites from which the **species** is known to have disappeared between 1999 and 2009, it is likely that the **species' distribution** is actually smaller than current **estimates**.
- They are vulnerable to **native predators** like foxes and **introduced species** like cats and dogs.
- Due to population **decline**, they are **classified** as a vulnerable **species** in Australia, with **estimates** of around 8,000 individuals.
- **Threats** include loss and modification of **habitat**, weed invasion, inappropriate fire and disturbance management, predation, competition from other **rodents** and potential climate change.

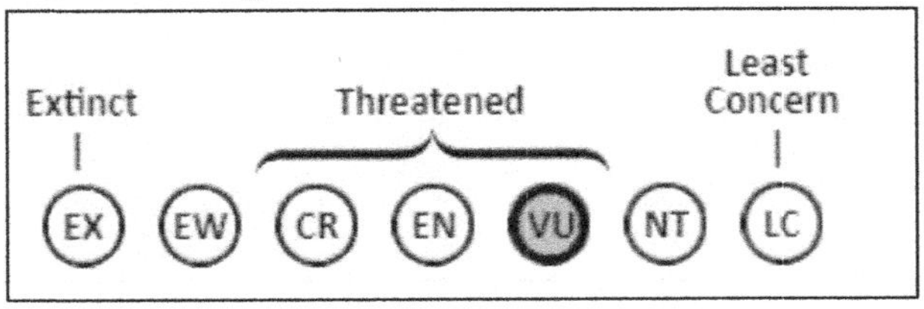

[1]

NCA List = V = Vulnerable

EPBC List = V = Vulnerable

Squatter pigeon (southern subspecies)

Geophaps scripta scripta

- Squatter pigeons are **unique** ground-dwelling birds with interesting features and behaviours. They are known for their distinctive black and white facial stripes, gray-brown **plumage**, and dark brown wings with some **iridescent** patches. The southern squatter pigeon is slightly larger than the northern (*Geophaps scripta*) **subspecies**, with blue-grey skin around its eyes, while the northern subspecies has orange-red skin.
- They are a medium-sized pigeon, typically around 30cm long and weighing between 190-250 grams.
- They also have a limited vocal repertoire, primarily using faint, mellow coos and plaintive crooning for communication.
- The southern squatter pigeon lives in arid and semi-arid areas, including grasslands and dry **sclerophyll forests**.
- Unlike many pigeons, squatter pigeons are primarily ground-dwelling birds, with a preference for **foraging** for food on the ground among sparse grasses.
- They feed on seeds, with some insects, particularly during the wet season. They **consume** seeds from both **native** and **introduced** plants, with grasses and legumes being particularly important.
- Squatter pigeons are reluctant fliers and often run away from danger rather than flying, displaying many sudden direction changes.
- They are usually seen in pairs or small **flocks**, but aggregations of over 30 have been reported.
- Squatter pigeons **roost** in low trees at night.
- Breeding in the squatter pigeon is poorly understood but appears to occur throughout the year and is likely influenced by **local** conditions. They breed as **solitary** pairs and may produce two **broods** in a season.

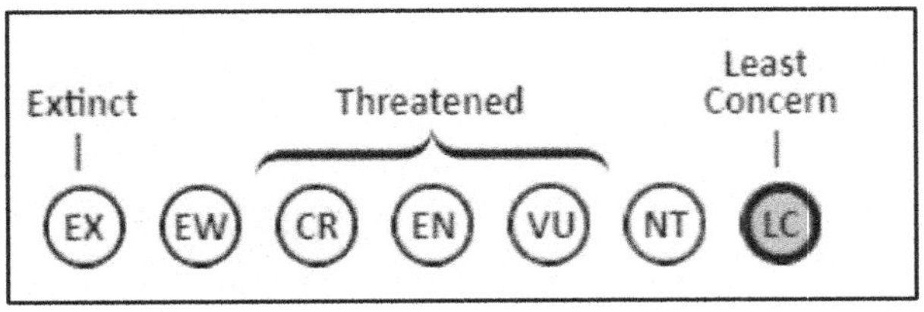

[1]

NCA List = V = Vulnerable
NCA NSW List= CR = Critically endangered
EPBC List = V = Vulnerable

Northern (*Geophaps scripta*) **subspecies**

Native guava

Rhodomyrtus psidioides

- The **native** guava, also known as *Bolwarra or copper laurel, is a **critically endangered** Australian **species** facing a significant **threat** from myrtle rust, a fungal **disease**. It's a shrubby tree, once common in **subtropical** rainforests, with fragrant, white or pink flowers that produce sweet, **edible**, urn-shaped berries.
- This **tropical** shrub can grow up to 5m in height and 3m in width
- *Bolwarra is the **Aboriginal** name for the **Native** Guava.
- Myrtle rust, a fungal **disease**, has decimated **native** guava **populations**, leading to their **critical** endangerment.
- **Native** guava is found in rainforest or wet forest areas along the eastern coast of Australia, from Gympie to Newcastle.
- It's a pioneer **species**, often found near **creeks** and **drainage** lines.
- The fleshy berries are sweet and creamy, ripen in winter, turning from green to brown and becoming soft to squeeze. They are ready to eat straight from the plant if you simply can't wait to try them. This aromatic fruit is often used as a spice-fruit in cooking, beverages, jellies, jams, and desserts. It can even be dried and crushed, then used as a spice.
- **Native** guava is known for its fragrant white or pink flowers, which bloom in summer.
- Aboriginal people used the plant's bark to make twine and fishing lines, and its wood to make spear shafts.
- Despite being a rainforest **native**, the plant is also drought-tolerant.
- The foliage is bright green in summer and can turn a coppery colour in winter and spring.
- Bolwarra is a **host food plant** for common red-eye butterflies and provides food for the weevils that **pollinate** its flowers.
- **Threats** to the **native** guava include **a decline in health/loss of mature plants,** and a lack of seed-based recruitment due to infection by myrtle rust, which has made the **native** guava one of the first Australian plants to be pushed to the brink of **extinction**.
- Other **threats** include **degradation** of **habitat** and competition from transformer weed **species**, clearing from rural, **agricultural** and urban development leading to edge effects, **degradation** and further **fragmentation**, **habitat degradation** and clearing due to forestry operations, too frequent/intense fire destroying **habitat** and individual plants, damage caused by inappropriate use of four-wheel drive vehicles, road and track development and maintenance.

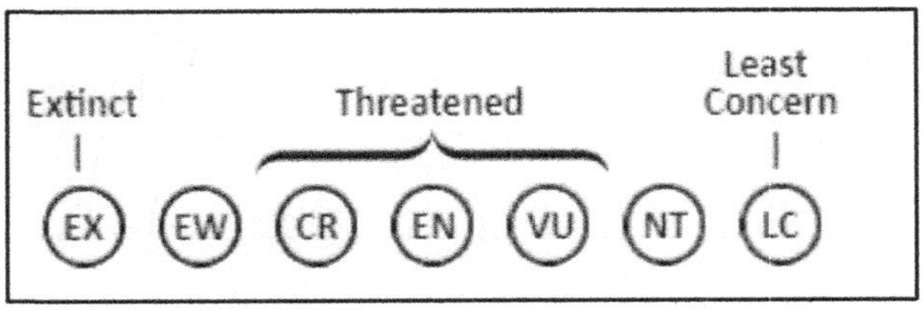

[1]

NCA List = CR = Critically endangered
EPBC List = CE = Critically endangered

Red and yellow mountain frog

Philoria kundagungan

- The red and yellow mountain frog is a small, brightly coloured frog found in the mountainous rainforests of Queensland and New South Wales.
- They are primarily found in **subtropical** and **temperate** rainforests, requiring high **moisture** levels.
- The mountain frog has a **restricted distribution** from the Mistake Mountains in south-east Queensland to Tooloom National Park, southwest of Woodenbong, in NSW. In NSW **populations** are known from Tooloom, Koreelah and Mount Clunie National Parks. In Queensland - a number of **populations** occur in Main Range National Park.
- They are characterised by their vibrant colours, limited range, and vulnerability to **habitat** loss and climate change.
- They are known for their striking appearance, with a dorsal surface that can be yellow, orange, red, or even black, and a bright yellow belly with a red throat. They have a wide head, a bluntly rounded **snout**, and **prominent** brown eyes. The skin is smooth, with the occasional low ridge or **tubercle** present in some **specimens**, which increases in frequency in individuals to the south of the **distribution range**.
- These frogs are relatively small, with a small, squat, pear-shaped body reaching a body length of up to 28mm.
- They are **rarely** seen above ground and are often hidden in shallow mud **burrows**.
- It can be difficult to find them because they are hard to hear and unclear when or under what conditions they call.
- Female frogs have large flanges on their first and second fingers, which they use to create a protective foam around their eggs.
- They breed on land, laying their eggs in water-filled cavities or nests. During the breeding season, males create small breeding chambers in wet areas, and females have specialised flanges on their fingers to help froth up protective foam around their eggs.
- Like many **amphibian**s, they are particularly sensitive to climate change, which can disrupt their breeding cycles and impact their **habitats**.
- They are only found in a few locations, with a few **populations** in New South Wales, like Tooloom National Park, Koorelah National Park, and southeast Queensland.
- **Threats** to the red and yellow mountain frog include risk of **local extinction** due to small, scattered **populations**, **isolation** of **populations** as a result of clearing and **fragmentation** of wet forest **remnants**, usually as a result of **agricultural** and forestry practices, **degradation** of **habitat**, infection by **amphibian chytrid fungus**, **habitat** disturbance and modification from **recreational 'herpers**' or frog enthusiasts targeting the **species**, potential **threat** of **feral** pigs damaging **species habitat** and climate change.
- Other **threats** include severe wildfires following drought (both likely results of climate change), which is shifting the habitat's climatic niche such that it can no longer sustain the species, particularly at low-elevation sites. It is considered likely that fire-related loss of leaf litter for shelter and soil moisture increases adult mortality and kills tadpoles.

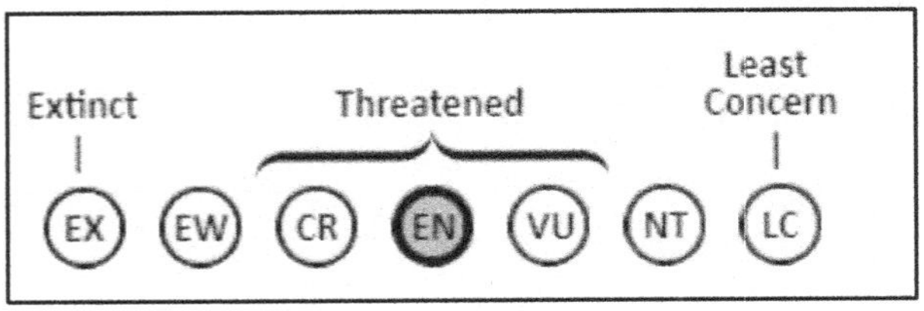

[1]

NCA List = E = Endangered
EPBC List = CE = Critically endangered

Eungella honeyeater

Bolemoreus hindwoodi

- The Eungella Honeyeater is a bird **species** found **specifically** in the Clarke Range near Eungella, Queensland. A relatively recent discovery, it was formally recognised as a distinct **species** in 1983. Interestingly, the **species** name "hindwoodi" honours amateur ornithologist, Keith Alfred Hindwood, who served as the President of the Royal Australasian Ornithologists Union.
- They gave it the common name of Eungella (pronounced yungella) honeyeater, after the small township of Eungella (an **Aboriginal** word meaning 'mountains of the mist'), which is perched on the edge of the Clarke Range escarpment.[25]
- The Eungella Honeyeater was one of the last new **species** of birds discovered in Australia.
- The Eungella Honeyeater was initially mistaken for a population of the bridled honeyeater.
- They primarily **inhabit** high-**altitude** rainforests, with most breeding activity occurring above 900 meters. They may visit lower **altitudes** during the non-breeding season, particularly around flowering eucalypts, casuarina forests, and woodlands.
- They are medium-sized honeyeaters with a grey overall appearance, slightly paler underparts, distinct markings on their cheeks, and a faint neck plume.
- Their **primary** food source is **nectar**, but they also **consume** insects, spiders, and fruit. Their role as a **pollinator** for various flowering plants is **significant**.
- The one recorded nest was a cup containing two eggs.
- Their call is a series of sharp, frantic-sounding whistles.
- The Eungella honeyeater faces several **threats**, primarily **habitat** loss and **degradation** due to logging and mining activities and potential climate change impacts. **Specific** concerns include **habitat destruction**, difficulty finding food sources, and potential disruption of nesting and breeding behaviours.

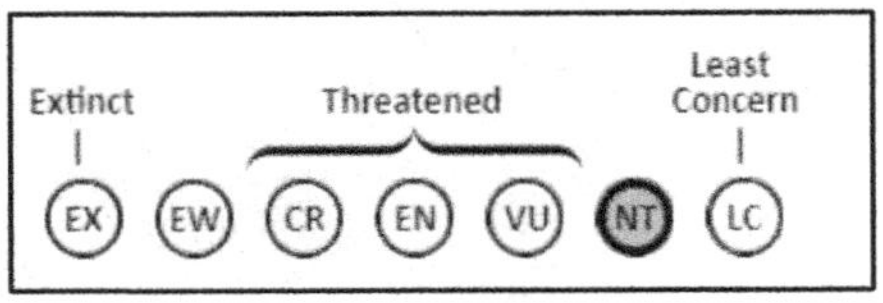

[1]

NCA List = V = Vulnerable
EPBC List = Not listed

Golden-tailed gecko

Strophurus taenicauda

- Golden-tailed geckos are **unique reptiles** known for their ability to squirt a foul-smelling, flammable fluid from their tails, a **defence mechanism** against **predators**. They also have the ability to detach their tails as a further escape strategy, with the tail **regenerating** later. These geckos are small, **nocturnal**, and **arboreal**, often found in forests and shrublands of eastern Australia.
- The golden-tailed gecko has almost perfectly round eyes, the colour of red marble, and a vertical slit-like iris. The eyes can also be orange or greyish.
- Geckos have some of the most incredible-looking eyes in the animal kingdom, with certain **species** displaying almost fantastical patterns across their **irises**, due to how their blood vessels show through.[26]
- Their vision is also 350 times better than a human's, and it gets better, because while we're stuck seeing nothing but shadows at night, geckos have **evolved** eyes that can see in full colour by the dim light of the Moon.[26]
- As adults, these geckos develop a golden or bright yellow stripe down their tail, contrasting wonderfully with the rest of their black and silver bodies.
- The **species** is **endemic** to Australia and occurs exclusively in Queensland, in the southern and central Brigalow regions.
- Golden-tailed geckos are just as happy up in the trees or shrubs as they are on the ground.
- When faced with would-be **predators**, they squirt a foul-smelling fluid up to 60cm from their position, from their tails, which can become highly flammable when mixed with ammonia.
- If a **predator** grabs their tail, they can detach it, and it will **regenerate** later.
- They typically measure about 71mm long, excluding the tail.
- They are **arboreal**, meaning they live in trees, and **nocturnal**, active at night.
- They **inhabit** forests and shrublands in eastern Australia.
- They often wait in **ambush** on trees, like cypress pines, to catch flying insects.
- They are **insectivores**, **consuming** insects like crickets, cockroaches, and moths.
- There are three **subspecies** of Golden-tailed gecko: *Strophurus taenicauda taenicauda, Strophurus taenicauda albiocularis* (northern golden-tailed gecko), and *Strophurus taenicauda triaureus* (central golden-tailed gecko).
- They can range in colour from black to orange to pale yellow.
- Courtship in this group is similar between **species** and involves the male licking then grasping the female over the shoulders and neck. Females lay small eggs, about 5 mm in length. Some **species** may produce multiple **clutches** every 3-6 weeks.

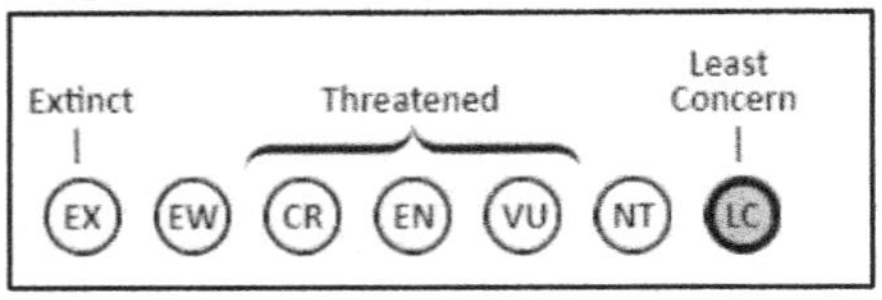

[1]

NCA List = NT = Near threatened
EPBC List = Not listed

Grimshaw's boronia

Boronia grimshawii

- Boronia is a **genus** of about 160 **species** of flowering plants. Most are **endemic** to Australia, with a few **species** in New Caledonia. It is found in understory shrub areas near gullies in wet open forests.
- Grimshaw's boronia is **native** to mountain ranges in central Queensland, growing 1.5 to 2.5m in height.
- Grimshaw's boronia appears to be able to **regenerate** well after disturbances, but more research is needed to see how it copes with prolonged or repeated disturbance.
- Boronias belong to the *Rutaceae* family, which includes citrus plants. This means they have oil glands in their leaves that can release a scent, and they can be fragrant or have a less pleasant scent, depending on the **species**.
- Boronias, including Grimshaw's Boronia, are known for their small, fragrant flowers, which can be various colours like pink, purple, and white.
- Plants in the **genus** *Boronia* are nearly always shrubs, although a very small number occur as herbs or as small trees.

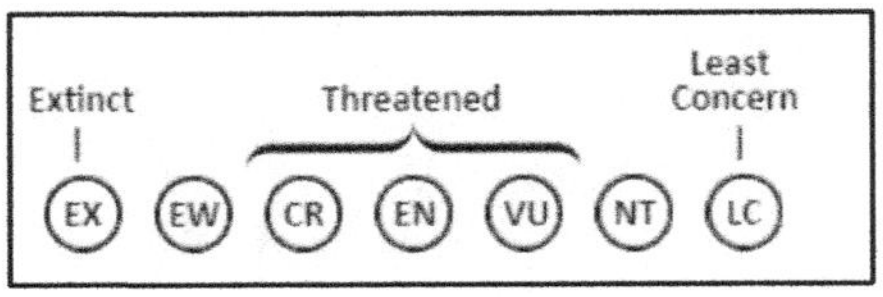

[1]

NCA List = V = Vulnerable
EPBC List = Not listed

Plumed frogmouth

Podargus ocellatus plumiferus

- Five subspecies of plumed frogmouth are recognised. This particular species is found in southeast Queensland. The Conondale ranges are considered to be a stronghold, with notable **populations** in Conondale National Park.
- The plumed frogmouth, also known as the marbled frogmouth, is a bird **species** found in eastern Australia, particularly in **subtropical** rainforests and wet **sclerophyll forests** at **altitudes of 50-800m**.
- The **species roosts** in the **canopy** of trees and is considered **cryptic** and extremely hard to find or study.
- They are **rare** birds, known for their distinctive "gobble gobble" call and **bill**-clapping.
- Masters of disguise, plumed frogmouths blend in so well with their surroundings that they look like tree branches, especially when resting during the day. Their mottled brown, black, and white feathers help them to look like tree bark or branches.
- Frogmouths are not great nest builders. But they think they are! The male bird will pick up a few sticks and leaves and shove them into place, bundling them together roughly in the shape of a nest. Frogmouth nests are flimsy, and it has been suggested that chicks are quite likely to fall. Most frogmouths seem oblivious to their failings as builders, re-using the same nest (or what remains of it), year after year.
- They are **nocturnal** birds, active at night and usually asleep during the day. They are active at night and are usually asleep during the day.
- When the weather is cold, frogmouths can go into **torpor** — a type of short **hibernation**.
- Plumed frogmouths are carnivores: they enjoy insects, worms, slugs, and even small lizards.
- While they might look a bit like owls, they are actually more closely related to nightjars, which are also **nocturnal** birds.
- Current **populations** are **threatened** by land clearing, inappropriate fire regimes and timber **harvesting,** with future impacts of climate change posing additional risks

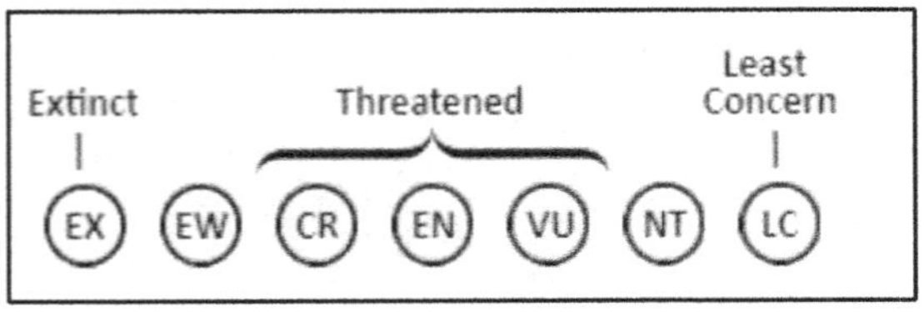

[1]

NCA List = V = Vulnerable
EPBC List = Not listed

Honey blue-eye

Pseudomugil mellis

NATIVE TO AUSTRALIA

- The honey blue-eye is generally found in clear lakes and lakes stained with **tannin**, as well as streams and wetlands, usually occurring where there is only a small amount of water movement and the fish can hide in the thick **aquatic vegetation,** such as **sedges** and along the margins of these bodies of water.
- They can be found as an isolated population in the Shoalwater Bay area, part of the Brigalow Belt north bioregion. This is a recently discovered **population** just south of Shoalwater Bay at the north end of Dismal Swamp. This **population** is independent of those found in the Wallum and on K'gari.
- A small number of fish are being bred and maintained in **captivity**, and the **species** is popular with people who keep **native** fish in **aquaria**.
- When feeding, the honey blue-eye forms **schools** of up to 30 individuals and moves through **vegetation** in search of food. They are **generalist feeders**, feeding on flies, insect **larvae,** and small **crustaceans,** and they also eat large quantities of **microscopic algae**.
- The honey blue-eye is a small, pale, honey-coloured fish that can grow up to 2.5–3cm, although the largest recorded size was a male fish at 3.8cm.
- As the common name indicates, the honey blue-eye has blue eyes, blue cheeks and a single thin line running along its body from the tip of the **pectoral** fin backwards. Males have distinct black bands and white margins on their fins.
- Honey blue-eyes form **schools** from about 25 to 70 individuals.
- **Threats** include competition from other **introduced** and **invasive** fish **species**, clearing and **degradation** of **habitat**, and collection for **aquariums**.

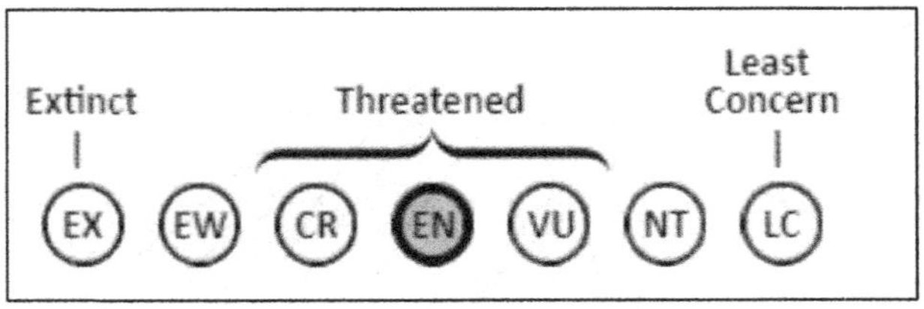

[1]

NCA List = E = Endangered

EPBC List = E = Endangered

Australasian bittern

Botaurus poiciloptilus

- The Australasian bittern is a large bird in the heron family, nicknamed the "**bunyip** bird".
- It is **secretive** and has a **distinctive** booming call, so it is more often heard than seen.
- It feeds on **frogs, eels**, freshwater **crustaceans,** and other **aquatic** animals.
- It is a **solitary** nester in **dense** wetland **vegetation** on **trampled** reeds and other ground-level plants.
- These birds are hard to find, so monitoring this **species** mainly relies on the ability to count the male birds based on their notable breeding calls. The best time to detect Australasian bitterns has been shown to be one hour before sunrise, in September (**Austral spring**), on a moonlit night with no clouds or rain.

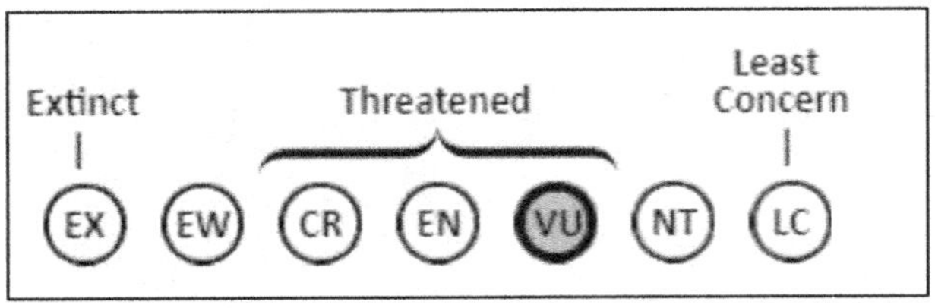

[1]

NCA List = E = Endangered

EPBC List = E = Endangered

❖ The **bunyip** is a **creature** from the **Aboriginal mythology** of southeastern Australia, said to lurk in **swamps, billabongs**, **creeks,** riverbeds, and **waterholes.**[3]

Swift parrot

Lathamus discolor

NATIVE TO AUSTRALIA

- The swift parrot is a slim, medium-sized parrot. As its name suggests, it has a very fast and direct flight. In flight, its body is **streamlined** with angular and pointed wings; it has a long, pointed purple-red tail. The body is mostly shades of very bright green, with a darker blue patch on the crown of its head. The forehead to the throat is bright crimson red, and there is a crimson patch at the bend of the wing. The male is slightly brighter in colour than the female.
- They are noisy, active and have a striking appearance. When feeding, the swift parrot is very **agile**, a bit of an **acrobat**, often hanging upside down.
- The swift parrot breeds in Tasmania and **migrates** to the mainland for winter, **foraging** on flowering eucalypts mostly in open ironbark forests.
- They feed in the **canopy** of flowering eucalypts, mainly eating **nectar**, but sometimes seeds and flowers, only coming down to drink.
- In dry years when flowering eucalypts do not produce enough flowers for them to feed on, swift parrots will travel greater distances to find food. Some birds may fly as far as coastal south-east Queensland.
- The swift parrot undertakes the longest parrot **migration** in the world, **migrating** between Tasmania and southeastern mainland Australia, covering distances up to 5,000 k!
- When in mainland Australia, they sometimes get together in big **flocks** and can be seen with smaller honeyeaters and lorikeets.
- They are usually **nomadic**, meaning they may spend only a few hours or weeks and sometimes months in some places. They **roost** at night as a group, often in the same trees.

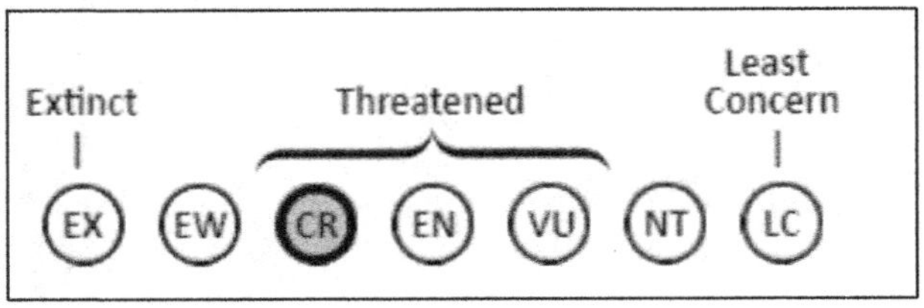

[1]

NCA List = E = Endangered (Qld)

EPBC List = CE = Critically endangered

Cascade treefrog

Litoria pearsoniana

NATIVE TO AUSTRALIA

- The cascade treefrog, also known as the Pearson's tree frog and Pearson's green tree frog, is a small tree frog growing up to 45mm in length. The mountain stream tree frog may be the same **species**, but the **specific** differences and similarities have not yet been thoroughly studied and are currently under review.
- The cascade treefrog's **dorsum** can range from completely light green to a mixture of green and brown, or mostly brown in colour. Usually, black dots are scattered over the **dorsal** surface, but in some **specimens**, these dots may not occur. The belly is white or off-white. A light brown/yellow stripe starts at the tip of the **snout**, runs across the nostril, eye, and **tympanum**, and then widens and **dissipates** over the shoulder. This line is underlined with a thicker, dark brown line. The **tympanum** of this **species** is brown, not green.[27]
- The tadpoles of the cascade treefrog are very similar in appearance to those of the **genus** *Ranoidea* and are difficult to tell apart; **range** is the best clue for **distinguishing** tadpoles.
- The **distribution** of the cascade treefrog is from just south of Gympie, Queensland, to the north of the Hunter Valley, NSW.
- This **species inhabits** flowing **creeks** in rainforests and **wet sclerophyll forests** in highland and coastal areas. Males call from **creek**-side **vegetation** during spring and summer. The call is an "eh-eh-eh-eh-eh-eh-ehk", with the sounds getting faster towards the end of the call, almost forming a trill.[27]
- The cascade treefrog has **declined** in certain areas in the past, likely due to the **amphibian chytrid fungus**. However, there has been a recovery in places, and it is currently considered the most widespread and **abundant** stream-dwelling frog in some wet forest areas.

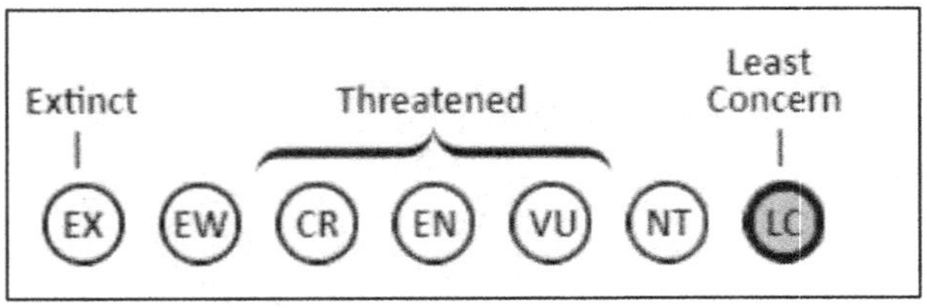

[1]

NCA List = V = Vulnerable

EPBC List = Not listed

***See page** for more general information about the physical **distinctions** to look for when trying to correctly identify frogs!*

White-bellied sea eagle

Haliaeetus leucogaster

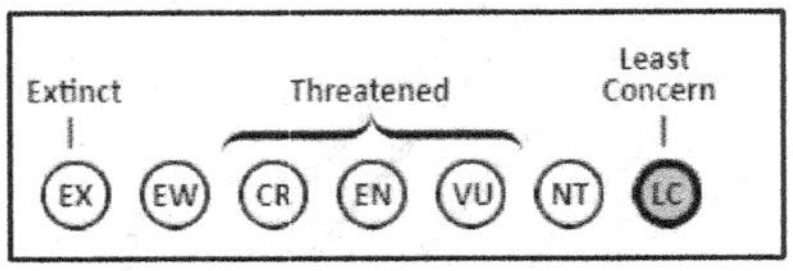

[1]

NCA List = not categorised
EPBC List = not categorised
Vulnerable in NSW

- The white-bellied sea eagle is a large, **diurnal raptor** with long, broad wings and a wedge-shaped tail. It measures 75–85cm in length and has a 180–220cm wingspan. Females weigh between 2.8kg and 4.2kg and are larger than the males, which weigh between 2.5kg and 3.7kg.
- The **plumage** of adult birds is mainly white and grey. The head, breast, and belly, and the feathering on the legs are white. The back and upper surfaces of the wings are grey, although the wings have black tips. The undersides of the wings are greyish-black around the **distal** edges, with a smaller area of white along the leading edge. The tail is grey at the base and has a white tip. The **bill** is bluish-grey with a blackish tip, the **iris** is dark brown, and the legs and feet are a cream colour. The **juveniles** differ from the adults in appearance, and there is a gradual transition from the brown and cream **plumage** of **juvenile** birds to the white and grey **plumage** of the adults across a series of **moults** over a period of several years.
- The white-bellied sea eagle is generally seen singly or in pairs, though it may occasionally **congregate** around sites where food is **abundant**. Its call is a loud, goose-like honking.
- The white-bellied sea eagle breeds and hunts near water, and fish form around half of its diet. It **opportunistically consumes carrion** and a wide variety of animals. They are skilled hunters that can attack **prey** as large as a swan. They also eat **carrion** like fish and sheep.
- **Human disturbance** to its **habitat** is the main **threat**, both from direct human activity near nests and the removal of suitable trees for nesting. The white-bellied sea eagle is **revered** by **indigenous people** in many parts of Australia and is the subject of various folk tales throughout its **range**.[28]

Eastern osprey

Pandion haliaetus cristatus

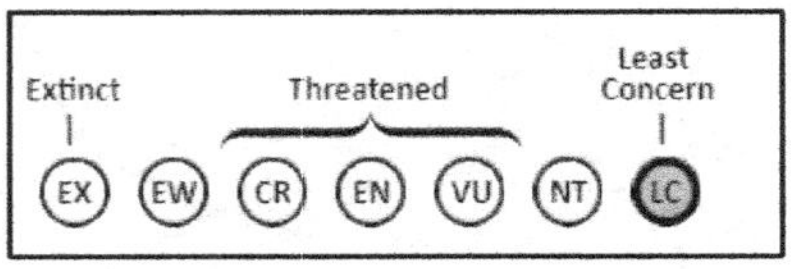

[1]

NCA List = not categorised
EPBC List = not categorised
Vulnerable in NSW

- An osprey may log more than 160,000 **migration** miles during its 15-to-20-year lifetime. Scientists track ospreys by strapping lightweight **satellite transmitters** to the birds' backs.[29] The oldest known osprey was at least 25 years and 2 months old.
- Ospreys are unusual among hawks in possessing a **reversible** outer toe that allows them to grasp with two toes in front and two behind. **Barbed** pads on the soles of the birds' feet help them grip slippery fish. When flying with **prey**, an osprey lines up its catch head-first for less wind resistance.[29]
- Ospreys are excellent **anglers**. Over several studies, the average time they spent hunting before making a catch was about 12 minutes.... something to think about next time you throw your line in the water!
- The osprey readily builds its nest on man-made structures, such as telephone poles, channel markers, **duck blinds**, and nest platforms designed especially for it. Such platforms have become an important tool in re-establishing ospreys in areas where they had disappeared. In some areas, nests are placed almost exclusively on **artificial** structures.[29]
- Osprey eggs do not **hatch** all at once. Instead, the first chick emerges up to five days before the last one. The older **hatchling** dominates its younger **siblings** and can **monopolise** the food the parents bring. If food is **abundant**, chicks share meals in relative harmony; younger ones may starve to death in times of **scarcity**.
- Their underwater eyesight is a superpower. Ospreys have a vision well **adapted** to detecting underwater objects from the air. **Prey** is first sighted when the eastern osprey is 10- 40m above the water, after which the bird hovers momentarily and then plunges, feet first with wings raised; a large splash is made as it enters the water.
- Ospreys are found near open water with an **abundant** supply of fish; listen for the osprey's whistling or chirping calls overhead, or look for this bird's **distinctive** flight profile and heavy wingbeats.[29].

Loggerhead turtle

Caretta caretta

- While present in various regions around the world, nine **distinct population segments** of loggerhead turtles are **recognised** under the Endangered **Species** Act. The South Pacific Ocean **population** nests mainly in Australia, with some nesting taking place in New Caledonia. In 1977, approximately 3,500 females were **estimated** to nest in this region. Currently, the **annual** number has **dwindled** to around 500, which is roughly one-seventh of the **population** observed over the previous 40 to 50 years.
- Loggerhead turtles are named for their large heads and strong jaws, which enable them to feed on hard-shelled **prey** like crabs, whelks, and conches.
- Adult loggerheads have a **carapace**, or top shell, that is somewhat heart-shaped and reddish-brown in colour. In contrast, their **plastron**, or bottom shell, tends to be a pale yellow.
- Unlike freshwater turtles and tortoises, sea turtles cannot retract their heads or flippers into their shells. **Hatchlings** are typically dark brown, with white or greyish-white edges on their flippers, and a generally yellowish to tan **plastron**.
- As **marine reptiles**, loggerhead turtles, like all sea turtles, must surface to breathe air. Adult females return to land to lay their eggs in the sand, showing quite amazing **navigation** skills by often returning to beaches near their **hatching** site from **decades** prior.
- Loggerhead turtles go through a **developmental life cycle** from **hatchling** to adult. After **hatching**, they spend the first 7 to 15 years in the open ocean before moving to coastal areas to **forage** and **mature**. Adults then undertake **extensive migrations** from their feeding grounds to their nesting beaches.
- **Satellite tracking** has revealed that Pacific loggerheads engage in trans-Pacific **migrations**. **Hatchlings** from nesting sites in Japan and Australia travel across the Pacific Ocean to feed near Baja California, Mexico, Peru, and Chile - an astonishing journey of nearly 8,000 miles. They spend up to 20 years growing to **maturity** before returning to the western Pacific Ocean to mate, nest, and complete their **life cycle**.
- In coastal waters, **juvenile** and adult loggerheads primarily feed on bottom-dwelling **invertebrates** such as whelks, other **molluscs**, horseshoe crabs, and various crabs, using their strong jaws to crush their **prey**.

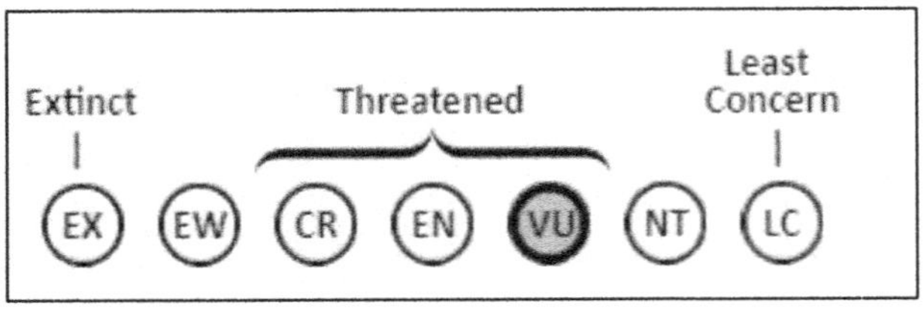

[1]

NCA List = E = Endangered

EPBC List = E = Endangered

Mouse bush

Homoranthus papillatus

- *Homoranthus papillatus*, commonly called mouse bush, is a **unique** Australian **native** plant known for its strong, distinctive, mouse-like odour when the leaves are crushed.
- It's a compact shrub with silvery blue foliage and clusters of small, pale yellow-green flowers that bloom in spring and summer.
- This plant is **endemic** to the Granite Belt of Queensland and has a limited **distribution** and **habitat**.

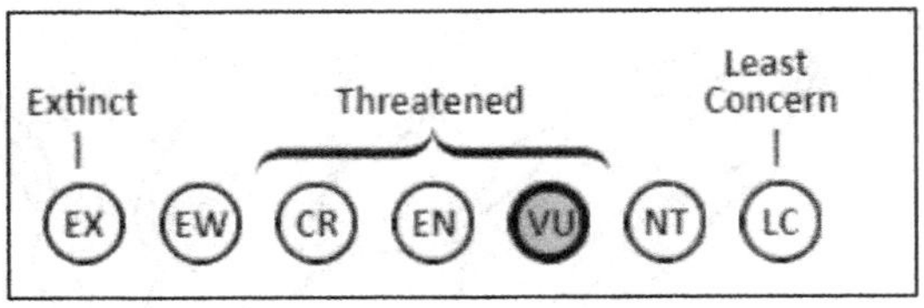

[1]

NCA List = V = Vulnerable

EPBC List = C = Critically endangered

Orange-speckled forest skink

Magmellia luteilateralis

- This skink is found only in a **specific** area of the Eungella rainforest, making it highly **susceptible** to **threats**.
- It prefers high-**altitude** rainforests, above 900m, which are more vulnerable to the effects of climate change. They prefer wet **sclerophyll forests** and **temperate** rainforests, particularly **moist** gullies and damp areas under rocks or logs.
- The **species** gets its name from the orange-speckled appearance of its **flanks**; the name *luteilateralis* refers to golden yellow on the **flanks**, though the **flanks** are more orange in reality.
- This **species** has a long body, widely spaced limbs and a pointed **snout**. Their colour is brown with dark and pale flecks on their scales and a sark line running along their side. Their belly is a light colour, ranging from cream to orange-yellow. They have tiny ear openings. **Snout** to base of tail length can be up to 5cm. The largest recorded **snout-to-vent** length is 92mm.
- Speckled forest skinks have a rough patch of scales on their ankles and a pocket of skin where the leg meets the body. The exact purpose of these scale features is unknown, but **parasitic** mites **congregate** in those areas.
- The skink spends time basking in patches of sunlight during the day, usually on and around large rotting logs, seeking cover when disturbed.
- They **forage** in leaf litter and around rocks and logs. They are **insectivores**, eating small **invertebrates**.
- They lay eggs.

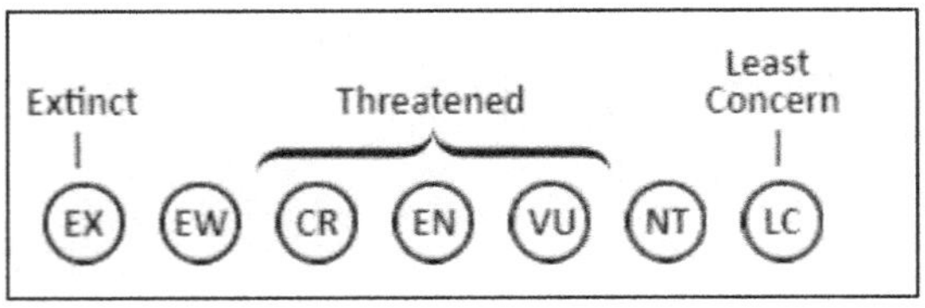

[1]

NCA List = V = Vulnerable

EPBC List = Not listed

Listed as Data Deficient on the IUCN Red List

Red goshawk

Erythrotriorchis radiatus

- Red goshawks are one of the **rarest** raptors in Australia.
- Red goshawks are large, swift raptors, known for their powerful builds and **unique** physical **characteristics**.
- They are primarily **predators** of medium to large birds, including **species** like kookaburras and cockatoos. They also **prey** on **mammals**, **reptiles** and large insects.
- They are known for their fast and powerful flight, with strong wing beats and glides.
- Thanks to its broad wings, a red goshawk can hunt at high speed, weaving in and out of trees, while its legs and talons can catch its **prey** in flight.
- They are **solitary** birds, generally silent and **inconspicuous**, often found in remote areas and near water sources.
- Females are **significantly** larger than males, making them one of the most sexually **dimorphic** raptors in the world. red goshawks are large hawks, with females reaching lengths of 60 cm and weighing up to 1200 g, while males are around 50 cm and weigh about 640 g. They have a wingspan of 111-136 cm.
- Females are nearly twice as heavy as males, a **characteristic** that makes them one of the most sexually **dimorphic** raptors in the world.
- They are known for their **rufous**-brown **plumage**, which is strongly streaked and barred, with large yellow legs and feet.
- The red goshawk has a distinctive underwing pattern of rust-red wing-lining contrasting with grey, heavily barred flight feathers.
- They prefer tall, open eucalypt forests and **riparian** areas, often nesting near permanent water sources in large trees. They build large stick nests in tall trees, often lined with twigs and leaves.
- Female goshawks **incubate** the eggs for up to 40 days, and the young stay with their parents for several months before dispersing.
- **Juvenile** red goshawks often make long-distance flights after leaving the nest, sometimes travelling over 1500km.

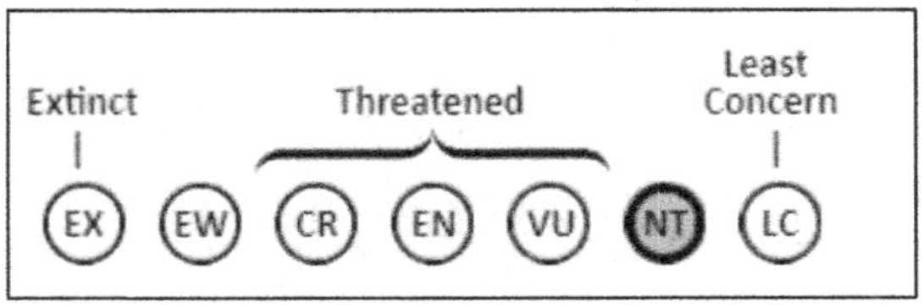

[1]

NCA List = E = Endangered

EPBC List = V = Vulnerable

Capricorn yellow chat

Epthianura crocea macgregori

- Yellow chats are small, lively, vibrant yellow birds with **unique adaptations** for **surviving** in harsh, **arid environments**. They weigh approximately 10g. The yellow colouring of their **plumage** may be an **adaptation** to reduce heat **absorption**, which helps them thermoregulate in their harsh **environment**.
- There are three **recognised subspecies** of yellow chats: the inland yellow chat (*Epthianura crocea crocea*), the Capricorn yellow chat (*Epthianura crocea macgregori*), and the Alligator Rivers Yellow Chat (*Epthianura crocea tunneyi*).
- They are sexually **dimorphic** - the adult male has a bright golden-yellow forehead and underparts with a **prominent** black crescent breast band.[30] The crown and nape are a grey-olive colour, the back is yellowish-brown, and the tail is black with yellow tips on the feathers.[30] The female is similar to the male but has a paler yellow colour and no breast band.
- They can also be identified by their **melodious** calls, including chirps and whistles, for communicating and **defending** their **territory**.
- Their **habitat** is subject to high temperatures for most of the year. It contains **marshy** plains vegetated with saltbush, rank grasses or cumbungi reeds (also known as bullrush – an **introduced**, **invasive species**). Most of these **habitats** are less than 5m above sea level on the threshold of the highest **annual** tides. These **marshes** have been formed by **marine** lagoons or inland artesian bore drains, creating a **saline environment**. This **distribution** includes the central **arid** region and the hot, sub**humid** monsoonal region. Most of the **annual** rainfall falls during the summer monsoon, leaving the remaining six months of the year with little rain.[30]
- Despite being categorised on the IUCN Red List as **least concern, the Capricorn yellow chat is **critically endangered**, with a wild population of less than 250 individuals. This is because of **habitat** loss, climate change, and other threats such as introduced predators, inappropriate grazing, and changes in water regimens. The Capricorn yellow chat was thought to be extinct until a small population was found in 1991.
- They have slower **metabolisms** and brush tongues, which may help them conserve or collect water. Some researchers suggest that their brush tongue may be an adaptation to aid in retrieving drinking water from dew or thin water films.
- They primarily eat **invertebrates**, including insects like ants, bees, and wasps.
- Yellow chats often occur in small groups of 2-10 individuals.
- The Capricorn yellow chat predominantly breeds in spring and summer, following rainfall and the freshwater inundation of the **marine** plains, which triggers food abundance. The nests are made from vegetation such as tiny twigs, rootlets and grasses. The female lays a **clutch** of two or three eggs, with incubation alternating between the male and female.

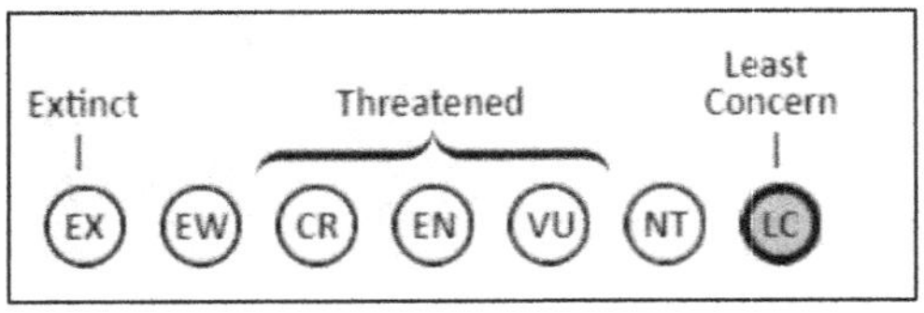

** see note

NCA List = CR = Critically endangered
EPBC = CE = Critically endangered

Blue-winged parrot

Neophema chrysostoma

- The blue-winged parrot is a small, striking Australian parrot known for its distinctive blue and yellow **plumage**.
- Blue-winged parrots are found in various **habitats** across south-eastern Australia and Tasmania, including grasslands, grassy woodlands, and areas near wetlands. They can also be seen in altered **environments** like golf courses and airfields.
- Their population size is considered to be relatively low in certain regions. In Queensland, they are listed as vulnerable on both the EPBC and NCA lists.
- Many aspects of the movements of the blue-winged parrot are poorly understood. There is a definite passage of blue-winged parrots to and from Tasmania after breeding each year, leaving in March to April and returning in August to October. Some birds, however, overwinter in Tasmania or on the Bass Strait islands.[31]
- Primarily ground **foragers**, they feed on the seeds of **native** and **introduced** grasses, herbs, and shrubs, usually in pairs or small parties
- They form **monogamous** pairs and build nests in hollows of trees or stumps, preferably one with a vertical opening. The eggs are laid on a bed of decayed wood. The **clutch** size is four to six eggs, with the female **incubating** the eggs and both parents feeding the nestlings. Interestingly, it is thought that the clearing of forests may have benefited this **species** in terms of feeding, but not with regard to breeding, as potential nest sites are removed.
- Males have a distinctive two-toned blue band across the face, while females are duller with a less **prominent** band. Both sexes have olive-green upperparts, blue wings, and yellow bellies.
- They are closely related to other small Australian parrots like the rock parrot and elegant parrot, with some having similar colourations.
- Their call is described as a high-pitched, chiming sound, like a "zit zit". They may also make soft, twittering sounds while feeding. They sound more like a thornbill or fairy wren than a parrot.

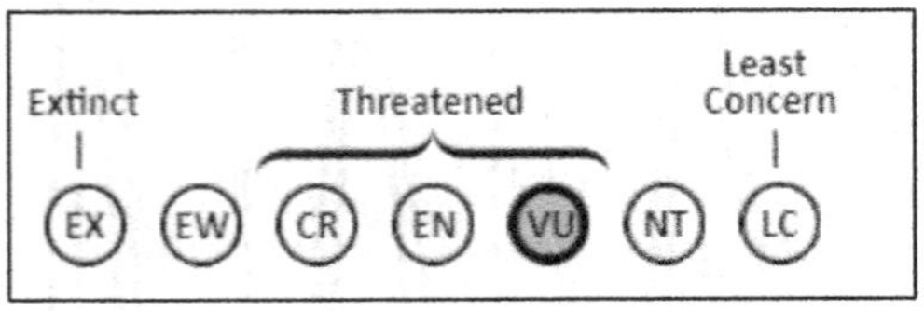

[1]

NCA List = V = Vulnerable
EPBC = V = Vulnerable

Narrow-leaved malletwood

Rhodamnia angustifolia

- Narrow-leaved malletwood is found in eastern Australian rainforests, particularly in areas with sand, alluvium, or volcanic soils.
- The **species** faces an extremely high risk of **extinction** in the wild.
- It thrives in various rainforest types, from sea level to adjacent ranges, and prefers certain soil types.
- It's found along the eastern coast of Australia, from New South Wales to Queensland.
- The tree has narrow leaves, a **dense canopy**, and can grow to 20m tall.
- The bark is soft, papery, and fissured, while the fruit is a berry that **matures** from green to black.

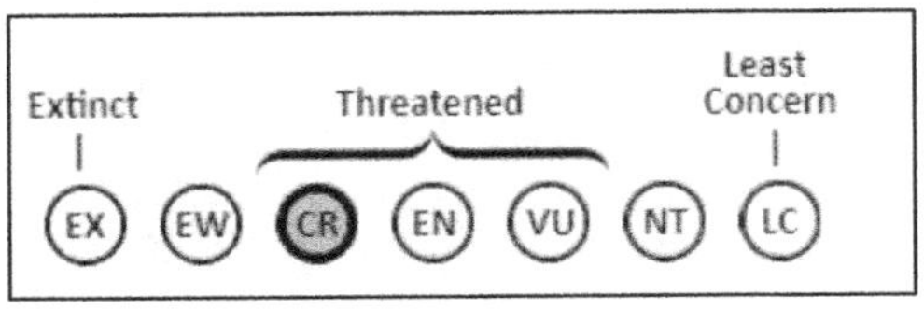

[1]

NCA List = CR = Critically endangered
EPBC = CE = Critically endangered

Mount Morgan myrtle

Decaspermum struckoilicum

- Mount Morgan, a town in Queensland, is known for its rich history as a gold and copper mining centre. The Mount Morgan Mine, which operated for 99 years, yielded **significant** amounts of gold, silver, and copper. The mine ultimately became a deep hole, measuring 1066 feet (325m) from the original mountain top.
- The Mount Morgan myrtle, also known as struck oil myrtle, is a plant **species native** to the area. It is known from five **localities**, about 8km east of Mount Morgan, in the area known as Struck Oil. It occurs in Bouldercombe Gorge Nature Reserve.
- Mount Morgan myrtle is an erect shrub or small tree, growing up to 4m high. It has elliptical leaves, white flowers borne in clusters, and dark, bluish-black, **globose** berries.
- While research is ongoing, myrtle rust is a plant **disease** that can impact various Australian **native species**, including Mount Morgan myrtle. However, some **native species**, including Mount Morgan myrtle, may have some degree of resistance to the **disease**.

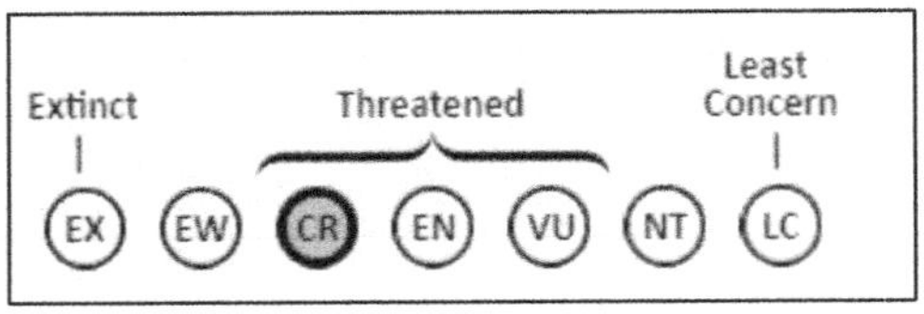

[1]

NCA List = CR = Critically endangered
EPBC = CE = Critically endangered

Star finch

Neochmia ruficauda ruficauda

- Star finches are a small grassfinch known for their bright red face and **bill**, and distinctive white spots. They are small birds with a wingspan of 49-56mm, a **bill** length of 11-13mm, and a weight of 10-12 grams.
- Star finches have a red face and **bill**, olive-green back and **flanks**, a cream to lemon belly, and numerous white spots on their head, breast, and **flanks**. The female has less crimson on the head and is generally duller than the male. The immature star finch is olive to brownish with a grey face and head.
- They have a stout, conical **bill adapted** for eating seeds.
- They are primarily seed-eaters but also **consume** insects and spiders.
- They are highly **social** birds, often seen in large **flocks** numbering between 10 and 30, but can build to hundreds of individuals
- Star finches are considered free breeders, meaning they will breed throughout the year. They are good parents with good fertility and a low mortality rate. They are considered a popular choice for breeding due to their hardiness and prolific breeding habits. During courtship, a female will carry a piece of grass in her beak, and the male will reciprocate by also holding a piece of grass.
- They are found in grasslands, grassy woodlands, and cleared areas near water sources. They are **granivorous**, **foraging** in vegetation just above the ground and often feeding on the ground, particularly in the dry season.
- While the star finch **species** overall is listed as least concern**, this **subspecies** is believed to number less than 50 individuals and is possibly extinct. The last reliable sighting was in 1994.[32]
- The **subspecies** *subclarescens* has three separate **subpopulations**; there may be 200,000 individuals, and the population appears to be **declining** in density.[32] The **subspecies** *clarescens has a stable population of about 3,500 individuals, but with a* patchy **distribution**.
- **Threats** include **habitat degradation** due to **overgrazing**, **invasive** weeds, and predation by **feral** cats.

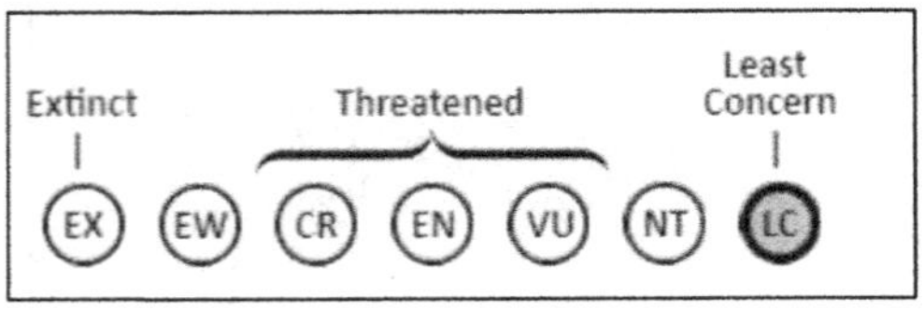

[1]

** see note

NCA List = E = Endangered
EPBC = E = Endangered

White throated needletail

Hirundapus australis

- White-throated needletails are large swifts with a **robust**, barrel-like body and long wings. The tail is short and square, with the protruding feather shafts giving a spiky, needle-like appearance.[33]
- They are greyish-brown except for a white throat and a white patch extending from the base of the tail to the **flanks**.[33]
- They can fly very fast - they are thought to reach speeds of 170km/h in **horizontal** flight! That is REALLY fast – <u>almost twice as fast as a cheetah</u>, which can reach speeds of 95km/h! The record for the fastest human running speed is 44.72km/h.
- White-throated needletails spend almost all their time in the air. For a time, it was commonly believed that they did not land while in Australia; it has now been observed that birds **roost** in **tree canopies** with **dense** leaves or in hollows, and **radio-tracking** has since confirmed that this is a regular activity.[34]
- White-throated needletails grow up to 20cm long and weigh up to 115-120g. They have a **lifespan** of about 8.5 years
- White-throated needletails breed in Central Asia and southern Siberia and **migrate** to winter south in the Indian Subcontinent, Southeast Asia, and Australia. They are **rare vagrants** in Western Europe and have been recorded as far west as Norway, Sweden, and Great Britain.
- In Australia, they are found in coastal areas of Queensland, NSW, and the western slopes of the Great Dividing Range; they also travel to Victoria, Tasmania and south-eastern South Australia.
- They **forage** over forests and grasslands in both mountainous and lowland areas, feeding on flying insects such as termites, ants, beetles, and flies. They catch the insects in flight in their wide beaks, usually feeding in rising **thermal currents** associated with **storm fronts** and bushfires. They are commonly seen moving with wind fronts; hence, they are also known as the storm bird! While feeding, they protect their eyes with a special **membrane** and a small ridge of feathers.
- They also drink while flying by **skimming** the surface of the water....they are able to do this because they have wide, gaping beaks.
- White-throated needletails communicate with each other using loud, sharp calls, chirps, a piping sound, and a churring twitter. They are also **social** birds and often form **flocks**.
- **Threats** to the White-throated Needletail include:
    - crashing into wind turbines, overhead wires, windows, and lighthouses
    - **insecticides** that reduce their food supply or cause secondary **poisoning**
    - loss of **habitat**, which means fewer places to roost and possibly less food
    - logging and hunting in the areas where it breeds in the Northern Hemisphere.[30]

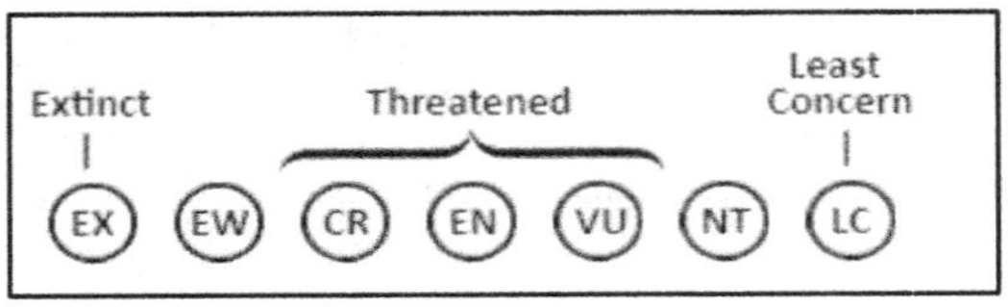

[1]

NCA List (NSW) = V = vulnerable
NCA List (ACT) = V = vulnerable
EPBC List = V = Vulnerable
Vulnerable in NSW

Grey-headed flying fox

Pteropus poliocephalus

NATIVE TO AUSTRALIA

- The grey-headed flying fox is amongst the largest bats in the world. They are commonly known as fruit bats or flying foxes and are also the largest bats in Australia.
- They eat fruits and other plant matter, occasionally also eating insects, which they can find due to their keen sense of smell.
- They are mostly **nocturnal**.
- Six flying fox **species** have already been made **extinct** (due to **overhunting**)!
- They are important **pollinators**.
- They can also carry **diseases** such as Australian bat Lyssavirus, rabies, Hendra virus, and Nipah virus.
- They have **cultural significance** to **indigenous** people, appearing in traditional art, folklore, and weaponry. Their fur and teeth were used as a form of **currency** in the past, and some **cultures** still use them as **currency**.
- **Variations** in body weight range from 120 – 1600g. Males are usually bigger than females.
- In terms of **threats**, flying foxes' **predators** include sea eagles and snakes.
- They also suffer from **habitat** loss where they can **forage**, competition, and **extreme weather events.** Recent studies indicate that since 1994, over 24,500 grey-headed flying foxes have died due to extreme heat events.

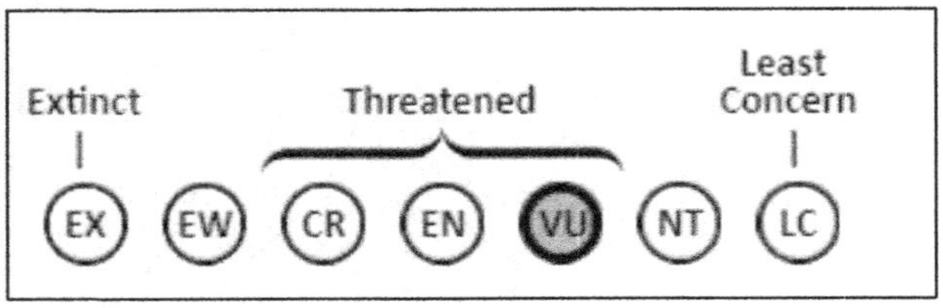

[1]

NCA List = LC = Least Concern

EPBC List = V = Vulnerable

Long-nosed potoroo

Potorous tridactylus tridactylus

NATIVE TO AUSTRALIA AND TASMANIA

- Potoroos are **marsupials** and are mainly **nocturnal**. They hide during the day in thick **vegetation** in forested areas and coastal **heaths**, but in the winter months, they might occasionally be out searching for food during the day.
- Potoroos prefer wet forests and wet scrublands with thick **understorey vegetation**, essential for cover. Eucalypt forests are important because potoroos rely on **fungi** associated with these trees.
- Potoroos are small members of the same family as bettongs and desert rat-kangaroos, looking a little like a bandicoot until they hop away with their front feet tucked tight to their chests.
- Five **species** in the *Potoroidae* family are **native** to Victoria: three bettong species and two potoroo **species**; sadly, all bettong **species** are **locally extinct**.
- Long-nosed potoroos are **solitary** in the wild, although when the females have young to care for, this is not so. They are not **territoria**l.
- The female potoroo carries nesting material with her tail, which is **semi-prehensile**.
- Their fur is brown-grey with a **rufous** tinge on **flanks** and pale grey underparts.
- They grow to have a head and body measurement of around 340 to 400mm and weigh 660 to 1600g. Their hind foot length is shorter than their head length.
- They can live up to 12 years in **captivity,** but average 5 to 6 years in the wild.
- **Fungi** form a large part of the potoroos' diet in the wild, and it is thought that they may play an important role in spreading **environmentally** important **fungi** within their **ranges**. Their diet also includes seeds, fruits, **vegetation**, **tubers**, and insects in the soil.
- Long-nosed potoroos give birth to one single youngster weighing 0.3g. The youngster stays in the pouch for between 120 and 130 days, stays near the mother after leaving the pouch, and still feeds from her for about another 40 days before becoming independent.
- **Threats** include the loss and **fragmentation** of natural **habitats**, which results in the **isolation** of **populations**, and **predation** by **introduced** cats and foxes.

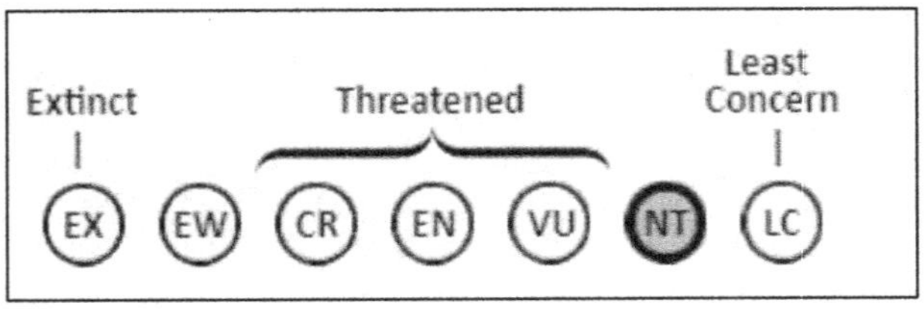

[1]

NCA List = V = Vulnerable

EPBC List = V = Vulnerable

Diamond firetail

Stagonopleura guttata

NATIVE TO AUSTRALIA

- The diamond firetail has also been referred to as the spotted finch, spotted grossbeak, spotted-sided grossbeak.
- It is a small, stocky bird but one of the largest finches in Australia, weighing 15–19g and having a wingspan of 64–71 mm.
- The birds are very **distinctive,** with a black breastband on a white breast. The **flanks** are black with white spots, and the bird has a scarlet **rump** and a black tail. The **bill** and eye ring are coral red, and the legs and feet are dark grey.
- Diamond firetails generally eat ripe or partially ripe seeds found on the ground. They occasionally eat insects and their **larvae**.
- During courtship, the male diamond firetail carries a long piece of green grass in his beak. Then, he sits or **perches** on a branch near the female, bobbing up and down. When the female approaches, he twists his neck around and opens his **bill**, just like a young bird begging for food.
- The diamond firetail **constructs** its nest using green grass blades and stems, lining it with fine grasses and feathers. Both partners build the nest, but only the female does the weaving. These nests can be found in trees and shrubs with **dense** foliage. Birdlife Australia records "To safeguard their eggs and nestlings, diamond firetails are often recorded building their nests into the base of the large stick nest of a bird of **prey**, such as a whistling kite, white-bellied sea-eagle, wedge-tailed eagle, brown falcon, nankeen kestrel or a square-tailed kite. One nest of a whistling kite contained nine diamond firetail nests!"[35] Brave birds to be game to do that!
- Living in open grassy eucalypt forest and farmland, woodland, mallee country, **heath** and grassland with scattered trees, these birds stay mostly in one place.
- The diamond firetail's **habitat** is under **threat** due to changes in **vegetation** structure caused by factors like **overgrazing** and weed **invasion**, leading to a loss of **primary** food plants and **habitat**. This situation increases competition with **invasive species** and the risk of **predation**.

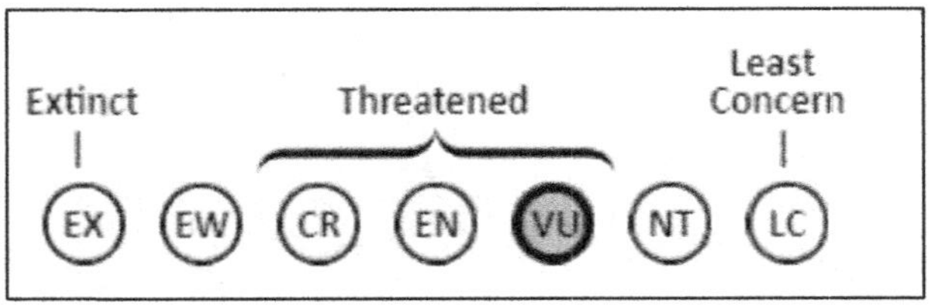

[1]

NCA List = V = Vulnerable

EPBC List = V = Vulnerable

No common name known

Apatophyllum flavovirens

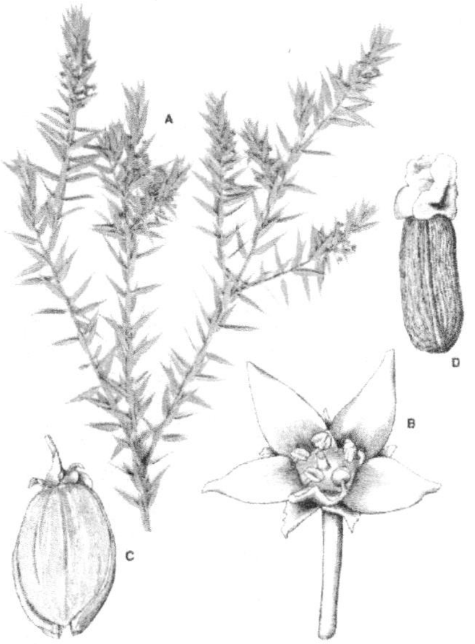

Distribution map for *Apatophyllum flavovirens*

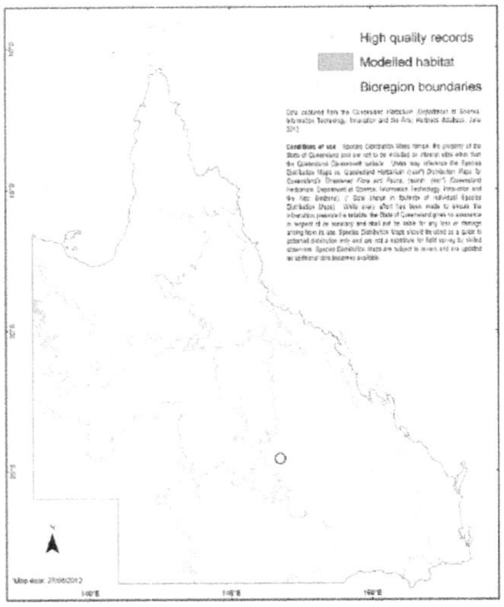

<u>There is a single recorded population of *Apatophyllum flavovirens*</u> with 19 plants at Bull Creek Gorge, Castlevale Nature Refuge, west of Carnarvon National Park (Forster, 2007; Queensland Herbarium, 2012) *Apatophyllum flavovirens* occurs as an **understorey** element in eucalypt dominated woodland on sandstone.

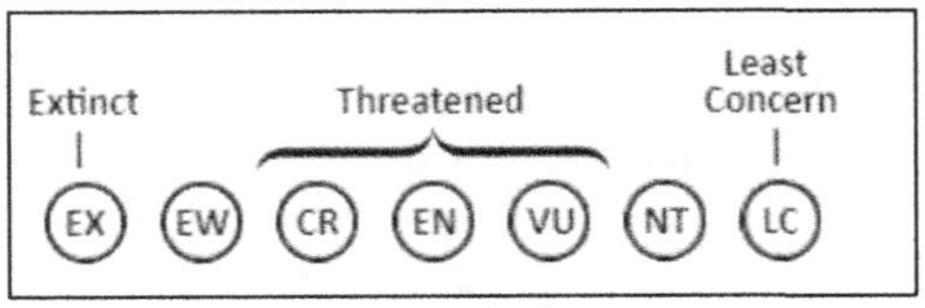

QLD NCA List = E = Endangered
EPBC List = Not listed

Threatening processes: The single known population of *Apatophyllum flavovirens* is not currently protected and is **threatened** by chance stochastic events (e.g., drought, wildfire) that reduce the population size or eliminate it entirely. Additional **threats** include **genetic** inbreeding depression as a result of the small population size (Forster, 2007).

No common name known

Commersonia beeronensis

Distribution map for *Commersonia beeronensis*

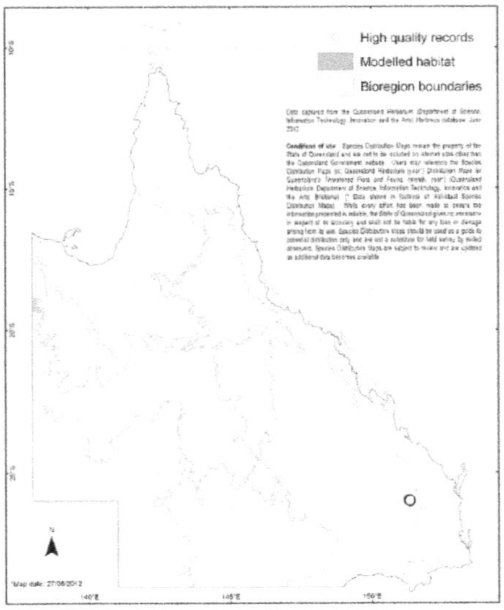

<u>*Commersonia beeronensis* is confined to Beeron National Park</u> (previously known as Beeron Holding) in south east Queensland (Guymer, 2005; Queensland Herbarium, 2012). *Commersonia beeronensis* occurs on granite outcrops, on slopes and at the base of granite domes in open eucalypt forests and woodlands dominated by *Eucalyptus dura*, *Corymbia petalophylla*, *E. acmenoides* and *Lophostemon confertus* (Guymer, 2005; Queensland Herbarium, 2012).

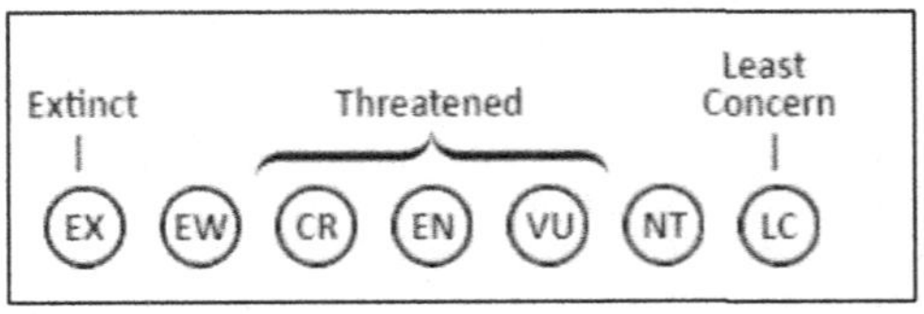

QLD NCA List = CR = Critically endangered
EPBC List = Not listed

Threatening processes: Commersonia beeronensis is only known from one area, and is an uncommon plant where it occurs. Severe wildfires kill the above ground parts of this plant with **regeneration** occurring from subterranean suckers and seeds (Guymer, 2005).

Queensland bottle tree

Brachychiton sp.

- The bottle tree is an interesting Australian **native** tree known for its swollen, bottle-like trunk that develops as it ages. The **characteristic** bottle-like trunk typically develops within 5-8 years of age.
- There are 30 **species** in Australia occurring in a variety of areas, but those in the Brigalow Belt are listed as **critically endangered**. In the Brigalow Belt, where much of the land has been cleared for **grazing**, these trees are protected and provide valuable shade for livestock.
- The swollen trunk results from the tree's ability to store water between the inner bark and the trunk. This **adaptation** allows it to survive long periods of drought. The tree also sheds its leaves during dry periods to conserve water. Bottle trees are highly drought-resistant and can thrive in **arid environments**.
- Bottle trees are known for being hardy and **adaptable**, **tolerating** a wide range of soil and climatic conditions. They can also be transplanted even when **mature**.
- **Indigenous** Australians have used the tree for various purposes, including food, water, and as a source of material for making rope and fishing nets. The tree holds **cultural** and spiritual significance for **indigenous** communities.

Powerful owl

Ninox strenua

- The powerful owl is the largest owl in Australia. Males are larger and stronger than females.
- Powerful owls are found in coastal forests in eastern Australia, usually not more than 200km inland, with a large home **range** of between 800 and 1000ha (although t**erritories** can be larger when food is **scarce**).
- They are **raptors** and are typically **territorial**.
- They like to live in tall, humid forests where their **prey** also lives.
- As an **apex predator**, the powerful owl mostly hunts **mammals** living in trees, particularly small to medium-sized **marsupials**. This **prey** provides about three-quarters of the powerful owl's diet.
- Like other owls, powerful owls have **evolved** to have some **unique adaptations.** They can hunt in almost total darkness due to the **unique anatomy** of their eyes, which have lots of light-sensitive cells. They also have remarkable hearing, allowing their brain to build almost a 3D space map to locate their **prey**.
- Owls cannot move their eyes up or down or from side to side? This is because their eyes are **cylindrical** and fixed in place, unlike human eyes! No wonder they always look as if they are looking straight at you! They are!!
- Owls have three eyelids: one for blinking, one for sleeping, and one for keeping their eye structures clean.
- Owls also have 14 **vertebrae** in their necks, allowing them up to 270 degrees of movement, meaning that they have a **significant** range of motion in their necks, and can look behind themselves or up and down! We humans only have seven **vertebrae** in our necks, so we have **significantly** less range of movement, up to a maximum of 180 degrees. Owls can achieve half again of what we humans can in terms of neck rotation!
- Owls also have poor day vision—they do not have as many 'rods' in their **retinae**—but despite their poor day vision, their night vision is similar to what we would see using night vision goggles!
- There can be some flexibility in terms of the **habitats** of powerful owls when they are not nesting.
- They have long intervals between the laying of eggs and the **hatching** of their **clutches**.
- The only **continent** in the world where owls are not found is Antarctica.

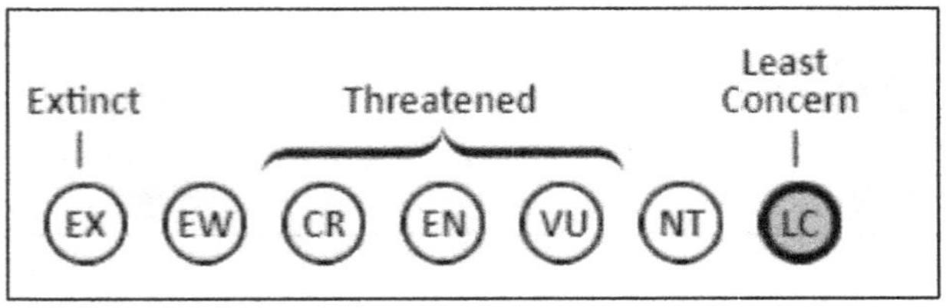

[1]

NCA List = V = Vulnerable
EPBC List = Not listed

Red-tailed tropicbird

Phaethon rubricauda

- The red-tailed tropicbird is the **rarest** of the tropicbirds and is **native** to the **tropical** parts of the Indian and Pacific Oceans. It is similar to a tern in appearance, with almost all-white **plumage**, a black mask, and a red **bill**.
- The bird has a **streamlined** but solid body shape, a wingspan of 111-119cm, a weight of 800g, and a length between 95-104cm, including the 35cm long red tail **streamers** in adult birds. These **tail streamers** are highly prized by the **Māori people** in the northernmost regions of New Zealand. Still, the birds are seen more frequently in ocean areas with water temperatures ranging from 23 to 30°C.
- Common names include red-tailed bos'nbird or silver bos'nbird and strawtail. The New Zealand Māori call it "amokura", and the **native** Hawaiians "koa'e'ula".
- The red-tailed tropicbird eats mainly squid and flying fish, which it catches by plunge-diving into the ocean or sometimes catching flying fish in the air.
- The red-tailed tropicbird nests in loose **colonies** on **oceanic** islands; the nest itself can be a scrape found on the face of cliffs, on a sandy beach or in a **crevice**.
- Only a single egg is laid and is **incubated** by both sexes for about six weeks. The parents make long trips **foraging** for food during the **incubation** period, sometimes for as much as six days.
- **Threats** include **habitat destruction**, pest introduction, and **predators** such as rats and **feral** cats that **prey** on eggs and young at nesting sites.
- Red-tailed tropicbirds are generally silent while flying. When leaving or arriving at the nest, birds may greet their mates with a squawk. The chicks chatter repetitively, as a begging call, whenever the parents are near, and the adult birds give a low, growling **defence** call.
- Strong winds can occasionally blow them inland, which explains sightings recorded away from the coastlines and their preferred **habitats**.
- The red-tailed tropicbird is strong in flight but walks with a shuffle on land with some difficulty, landing by flying into the wind, then **stalling** and dropping to the ground. Nests are usually no more than a metre from the edge of the shaded area or scrub, so it does not have to walk too far.
- They can hover in midair by flying into the wind; pairs may even fly backward and circle each other during courtship. They can be **territorial,** with **bill**-jabbing and fights breaking out, the two **combatants** locking **bills** and wrestling for up to 90 minutes.

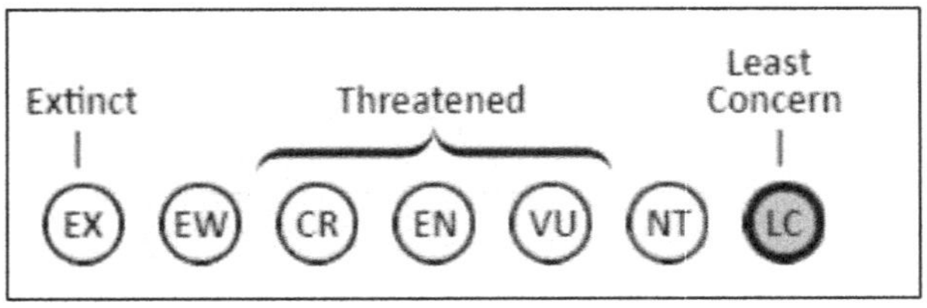

[1]

NCA List = V = Vulnerable

EPBC List = Not listed

Western Alaskan bar-tailed godwit

Limosa lapponica baueri

- Bar-tailed godwits are large wading shorebirds with a length of 37-39cm, a wingspan of 62.75cm, and a weight between 250-450g. They have a long neck and a long, upturned **bill** with a dark tip.
- Bar-tailed godwits do not breed in Australia but come here from Scandinavia, northern Asia, and Alaska for the winter. **Satellite tracking** has confirmed that some birds fly over 11,000km, and one tracked godwit flew 29280km in a round trip lasting 174 days!
- The western Alaskan bar-tailed godwit **migration** across the Pacific Ocean is the longest known non-stop flight of any bird and the longest journey without pausing to feed by any animal.[36]
- Their breeding **plumage** is a distinctive rusty red colour.
- They feed on bristleworms and **shellfish** on coastal mudflats, beaches, mangroves, and estuaries, **probing** their long **bills rapidly** into the shallows or exposed mud.
- Feeding parties can be 30 or more birds, including **migratory** adults and some birds that remain all year round, mostly young birds.
- **Threats** include **habitat** loss and **degradation**, **overexploitation** of **shellfish**, **pollution/contamination** impacts, **diseases**, **extreme weather events**, and climate change, which can raise sea levels.

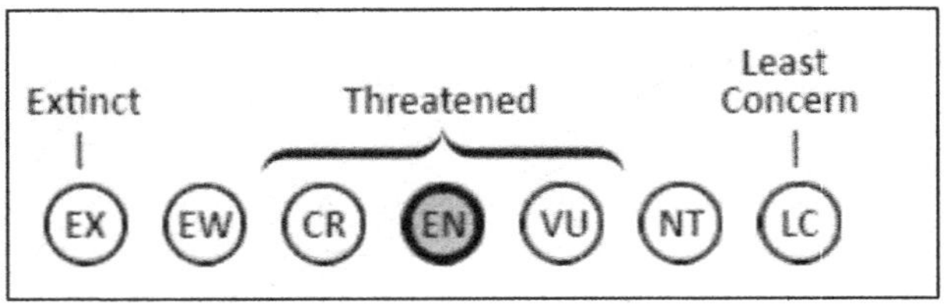

[1]

NCA List = V = Vulnerable

EPBC list = E = Endangered

Mount Elliot broad-tailed gecko

Phyllurus amnicola

- The Mount Elliot broad-tailed gecko is an interesting **reptile** due to its limited **range**, **unique habitat**, and **camouflage adaptations**.
- This gecko **species** is only found on Mount Elliot in Bowling Green Bay National Park, Queensland, living among boulders and along **creek** lines. It is considered **microendemic**, meaning it has a very small natural **range**.
- They are **specifically** found near streams and **creeks**, usually on rocks or boulders just above the water line, with their head towards the water.
- "Amnicola" translates from Latin to English as "dwelling by a river".[37]
- They are beige or white in colour with dark splotches and/or stripes running crosswise on their bodies. Their limbs are usually banded with the base colour getting lighter as it proceeds **ventrally**, with the bands becoming more frequent.
- Their mottled colour helps them blend in with bark or rocks, and they have low, bumpy **tubercles** over the body.
- Their limbs are also long and have large, **distally** compressed **digits**. The tail, which resembles a leaf and has a long knob-like tip, is compressed, as is the body's morphology. This compressed **morphology** allows the gecko to lie flat against boulders and rocks.
- Their large, fleshy tails are not only used for **defence** but also for storing fat.
- They can drop their tail as a decoy if **threatened**, and it is noted that regenerated tails are chunkier, mottled and smooth, usually not as bright as the original tails, which are the same colour as the body.
- They have a total length of about 15cm, with a **snout-to-vent** length of 9.5cm.
- They **forage** near turbulent or flowing water, which is thought to be why the geckos are always found in this location and the head-down position.
- They are **nocturnal ambush hunters**, relying on **camouflage** and patience to catch **prey** and primarily eat large **nocturnal** insects like spiders, moths, cockroaches, and beetles.

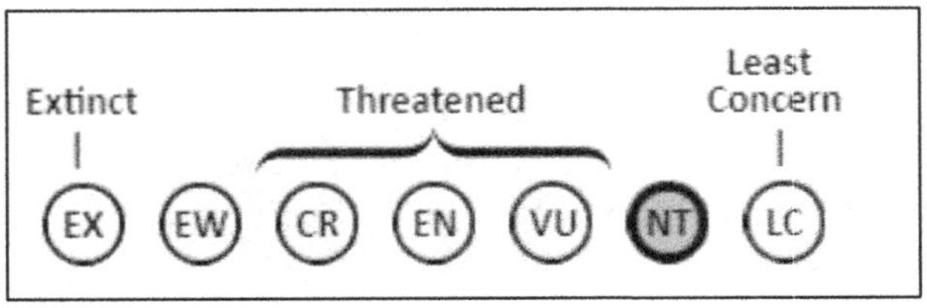

[1]

NCA List = V = Vulnerable

EPBC list = E = Endangered

Northern bettong

Bettongia tropica

- Northern Bettongs are Australian **marsupials** with **unique** habits! Like kangaroos, they carry their babies in a pouch.
- Their average body length is just over 30cm, and they weigh around 1kg, about the size of a rabbit. All bettongs have long tails, roughly equal to body length.
- They're like little diggers, using their front claws to unearth underground **fungi**.
- A **solitary** and **nocturnal** animal, Northern Bettong spend the day sheltering in a nest of grass, leaves and bark, emerging at night to **forage**
- They are fussy eaters and mainly eat **truffles**. They also eat **tubers**, herbs, fruits and seeds, particularly in drier times.
- They also use their long tails like a fifth limb, helping them dig and build nests.
- These little animals play an important role in the forest, helping spread **truffle spores,** which are important for tree health.
- They are a **keystone species** that helps disperse **ectomycorrhizal fungi**, which form a **symbiotic** relationship with eucalyptus trees. Without the bettongs, the whole **ecosystem** could collapse – lose the bettongs and the **truffle spores** may not be so effectively dispersed, as the bettongs are one of the top **truffle** dispersers. This, in turn, would impact the trees, which rely on the **truffles** to fix and make **nutrients** available from the very poor soils that they are growing in.
- They move around with a low, springy hop and can move at speed if startled.
- They leave behind **undigested** grass "oorts" that help researchers track them.
- The northern bettong builds a private nest of leaves, grass, and bark. They use their **prehensile** tail to carry nesting materials. The northern bettong can **reproduce** year-round, and a female can produce up to three young per year. The young stay in the pouch for about 100 days.
- They typically live for four to six years.
- Once widespread, they are **rapidly declining**. It is estimated that only 1000 are left in the wild, with only two known **populations**.
- The northern bettong faces many **threats**, including **habitat** loss, competition with pigs, and climate change.
- The most **significant threat** to the northern bettong is **habitat** loss caused by land clearing for **agriculture.**[38]

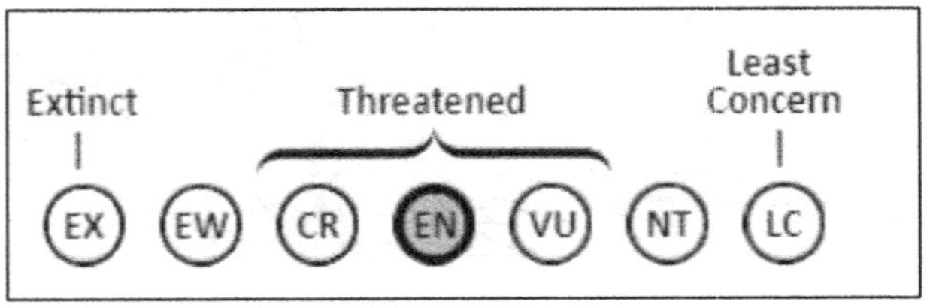

NCA List = E = Endangered

EPBC list = E = Endangered

Bunya pine

Araucaria bidwillii

© Axel Aucouturier

This evergreen, fast-growing conifer can reach up to 30-50m tall and is considered a living fossil from the Jurassic; bunya pines live for around 300 years.

Its natural habitat is fertile basalt soils in cool and moist valleys at low elevations near the coast. It is normally found as an emergent tree over tropical rainforests in humid areas where mists are frequent.

The trunk is straight and undivided, sometimes free of branches for up to two-thirds of the tree's height with very little taper.

Araucaria bidwillii have two types of foliage: the younger or understory needles are narrow, glossy green with a point and are arranged in two rows, while mature leaves are stiff with very sharp points and spirally arranged around the branchlets. Bunyas produce very large, spiny female cones up to 60cm long and weighing 5-6kg! The cones take two years to mature, each containing up to a hundred seeds.

These cones were a very important food source for native Australians. Each Aboriginal family owned a group of trees, which were passed down from generation to generation. This is said to be the only case of hereditary personal property owned by the Aborigines. Tribal wars were stopped during the season to organise festivals, climb 50m tall bunyas, and harvest the cones.

The trees were then greatly over-exploited by the first European settlers, who felled the vast majority for their timber. Bunya was highly valued as "tonewood" for stringed instruments' sound boards.

Few remaining wild stands are now protected in reserves and National Parks and if few subpopulations are still currently threatened in northern Queensland, the main in the South are not.

Bunya pines in the Bunya Mountains National Park (which is on the border of the Brigalow Belt) are being infected by *Phytophthora multivora*, a microscopic water mould that causes dieback. This disease is different from the one affecting bunya pines in the Maleny area (which is not within the Brigalow Belt), which is caused by *Phytophthora cinnamomi*. *Phytophthora* is a soil-borne pathogen that attacks the roots of trees, preventing them from absorbing water and nutrients, ultimately leading to death.

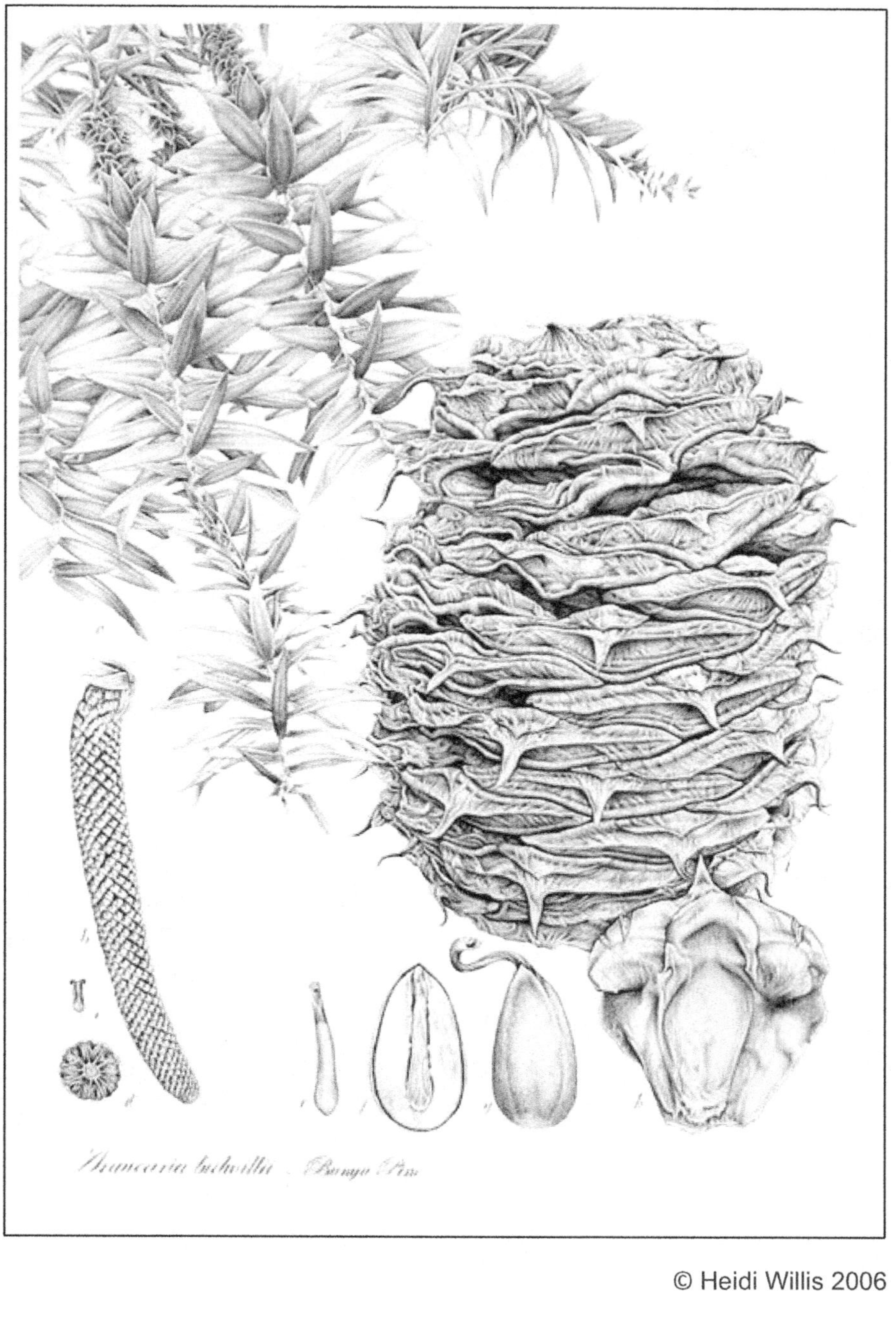

Currently, management efforts focus on limiting the spread of infected soil through measures like boot-washing stations at track entrances and promoting awareness about responsible park usage.

<u>Bunya Mountains National Park is home to the largest stand of bunya pines remaining in the world and is culturally significant.</u>

Just for fun: fins and feathers!

Bony fish anatomy:[37]

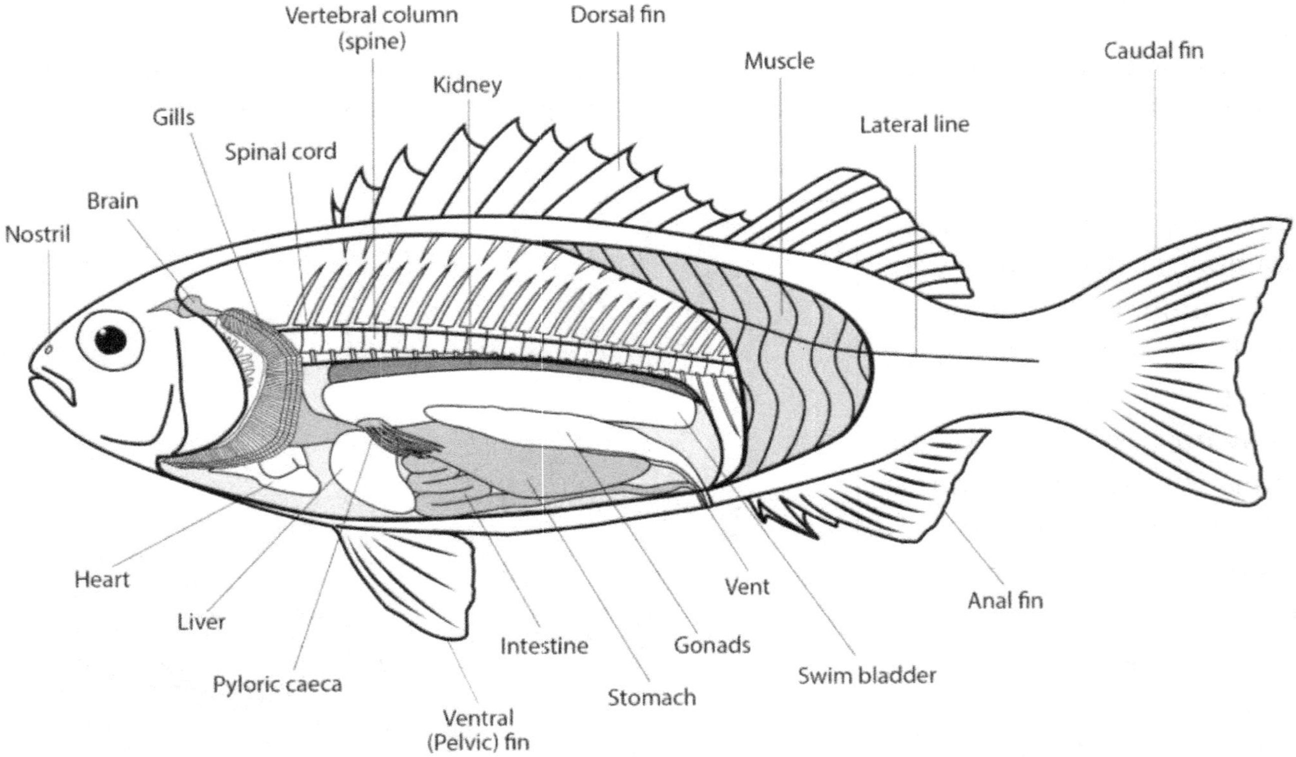

Basic bird anatomy:[38]

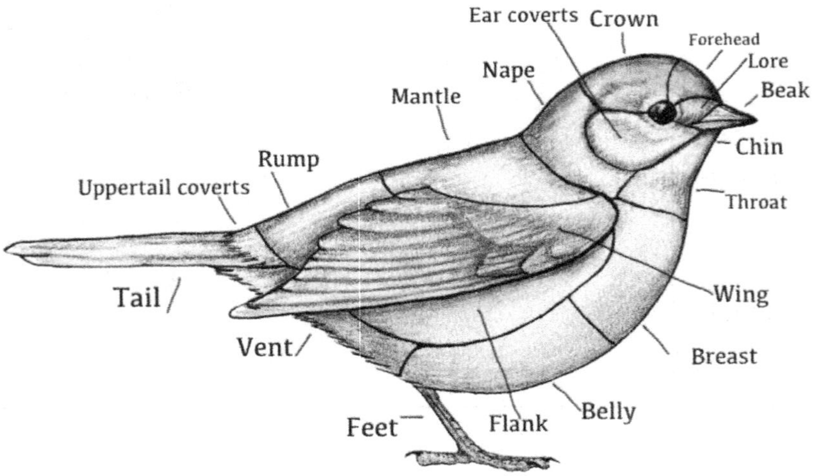

GLOSSARY – All those big or unusual words! ...

Some of these words are really unfamiliar, even to me as the author of this book, but I have included them because I think they could be useful for any person, child or adult, who has an interest in nature. For the most part, a synonym (word that means the same thing) has been used to try to keep these explanations as simple as possible. Definitions and synonyms from Dictionary.com and Thesaurus.com have been used on occasion. These definitions are not intended to be 100% perfect, but to give younger readers (and perhaps the adults reading with them) some other ideas to help them understand and explain what may be unfamiliar words, in the context of this book! Not all terms will be found in the text; however, they are still good terms to learn for any enthusiastic nature lover.

---

abdomen: belly, gut, midsection
aboriginal: original, earliest, native
aboriginal mythology: The "Dreaming" - the rich oral traditions, stories, and beliefs of indigenous Australians, passed down through generations
absorb: take in, soak up
abundant: ample, plenty of, extensive
abyssal (ocean): relating to or denoting the **depths** or bed of the ocean, especially between about 3000 and 6000m down
abyssopelagic (ocean zone): abyssal zone
achene (botany): a small, dry one-seeded fruit that does not open to release the seed
acid, acidic: substance like lemon, sour
acrobat, acrobatic/s: like gymnastics, athletic
adapt, adapted: change to suit, alter
adaptation/s: adjustment/s, alteration
adhere, adhesive: stick, sticky like glue
adjoin/s, adjoining: be next to and joined with
aggressive: ready or likely to attack or confront
agile: athletic, quick, energetic
agriculture, agricultural: the science or practice of farming
akin to: similar nature or character; related by blood
albino, albinism: pale or white colouring, usually with pink eyes
alga, algae: plants living in water
altitude, altitudes: height in relation to sea or ground level
ambush: trick, trap, a surprise attack
ambush hunter, ambush predator: an animal that prefers to capture or trap prey by stealth or strategy
amphibian: an animal that can live both on land and in water, cold-blooded vertebrate animal
anal: related to or near the anus
analysis: study, test, investigate
anatomy/anatomical: structure of a plant or animal, the way it is built
angler: an animal/person who catches fish
annual, annually: yearly, once a year
antenna, antennae: feelers
antennules: small antenna, especially one of the foremost pair of a crustacean
anterior (anatomy): towards the front
anther: the part of the stamen of a flower that produces and contains pollen and is usually borne on a stalk
antivenom: a substance used to neutralise or counteract the effects of a venom
anus: the opening at the end of the digestive system
apex predator: a meat eating animal that preys on other animals but has no natural predators of its own, at the top of the food chain
aquarium, aquaria: fishbowl, fish tank
aquatic: occurring in water
aquatic vegetation: plants that rely on wet environments
arboreal: living in or among trees

archaea: **microorganisms** which are similar to **bacteria** in size and simplicity of structure, but radically different in molecular organisation. They are now believed to constitute an ancient group which is intermediate between the **bacteria** and eukaryotes

arid (of land or a climate): having little or no rain; too dry or **barren** to support **vegetation**

arthropod: an animal with a segmented body, jointed limbs and usually a shell including insects, spiders, and crustaceans

artificial: not natural, made by humans

assimilate: ingest/digest, soak up

Austral spring: a season of spring in the Southern hemisphere where the The greatest amount of ozone is lost, generally beginning in September

awn: a stiff **bristle**, especially one of those growing from the ear or flower of **barley**, rye, and many grasses

axis: an imaginary line that goes through the middle of something

bacteria/bacterium: a member of a large group of **unicellular microorganisms** which have cell walls but lack **organelles** and an organised nucleus, including some that can cause disease.

bacterivore/bacterivorous: an organism which obtains energy and nutrients primarily or entirely from the consumption of **bacteria**.

barb: a sharp projection near the end of an arrow, fish hook, or similar object, which is **angled** away from the main point so as to make extraction difficult; a fleshy filament at the mouth of some fish such as **barbel** and **catfish**.

barbed: having a barb or barbs

barbel: a **fleshy filament** growing at the mouth of some fish

bathypelagic (ocean zone): midnight zone

bill (of birds): beak; the hard, pointy part of a bird's mouth

billabong/s: small lake usually formed when the path of a creek or river changes leaving a dead end

biocide/s: a substance that kills life

biodiversity: differences in biology within plant and animal species

bite force: force applied between the upper and lower mouth structures, jaw and/or teeth

bivalve: any mollusc with two shells hinged together

botany: the scientific study of the physiology, structure, genetics, ecology, distribution, classification, and economic importance of plants

bracken: type of big, bushy, green fern

brackish: water that is saltier than fresh water but not as salty as seawater

brood: a number of offspring, clutch, youngsters

brood pouch: a pouch where eggs develop and hatch

bunyip: the bunyip is a creature from the aboriginal mythology of southeastern Australia, said to lurk in swamps, billabongs, creeks, riverbeds, and waterholes

burrow: hole or tunnel dug by an animal, usually for a home or hiding place

buttress roots: large, wide roots sometimes known as plan roots on all sides of a tree

calligraphy: decorative handwriting or handwritten lettering

camouflage: disguise, mask, concealment

canine/s: a pointed tooth between the **incisors** and **premolars** of a mammal, often greatly enlarged in **carnivores**

canopy: the upper part of a tree or plant group that's made up of leaves and branches. It's also known as the overstory

captive breeding: bred in captivity

captivity: kept in a place like a zoo or a cage

carapace: a bony or chitinous shell, test or shield covering some or all of the dorsal part of an animal, e.g. turtle

carnivore/carnivorous: meat eating

carotenoid: nature's food colouring! They're the bright yellow, orange, and red pigments that give fruits and vegetables their colourful appearance

carpal: relating to the bones forming the wrist

carrion: dead and rotting flesh of an animal

cartilage: strong, bendy, and rubbery tissue

cartilaginous tissue: made out of cartilage like our ears and nose tips

caudal (anatomy): of, relating to, or being a tail; directed toward or situated in or near the tail or posterior part of the body

cauline (botany): relating to or growing from a plant stem, particularly the upper part of the stem

characteristic: distinctive, singular feature

chemical cues: hints that are chemical, could be particular smell

chenopod grassland: *Chenopodioideae* are annual or perennial herbs, subshrubs, shrub or small trees

chromatophore: tiny colour-changing cells under an animal's skin

chrysalis: the hard-shelled pupa of a moth or butterfly

chytrid fungus: an infectious fungal disease which can be fatal to amphibians

chytridiomycosis: is an **infectious disease** of **amphibians** – see **chytrid fungus**

circulate: move in a circle, pass from place to place, spread

circumference: the distance/measurement all the way around an object/circle

cite, cited: quote, name, reference

classify, classification (scientific): the sorting of things into groups based on how they are alike

cloaca/cloacal: a shared exit for bodily waste and reproductive function

clutch (of eggs): total number of eggs/young in one nesting

coastal dredging: excavating material from a coastal water environment

co-exist: live or exist together

colony/colonies: group, settlement, swarm

combatant/s: fighter, battler

commercial: merchant, business, of trade, public

commercial port: a place for ships to load or unload things for buying and selling

complex: complicated, convoluted

cones (eyes): specialised cells that allow us to see colour and fine details, particularly in bright light

configuration: composition, shape, structure

congregate: gather into a crowd or mass

conservation: saving and protecting

conspicuous: easily seen or noticed; standing out from a group

construct: build or put together

consume: eat, drink or use something up

contamination: infection, pollution, corruption

continent: big, solid area of land – Earth has seven continents

co-operative: working with others to help make things happen

coronal (anatomy): dividing body into front and back

counterpart: equals, fellow, like, twin

courting, courtship: set of display behaviours in which an animal, usually a male, attempts to attract a mate

cranial (anatomy): towards the head

creature: a living animal or living insect

creche: a group of young cared for communally

credible: believable, reasonable, likely
creek: small stream or river
crepuscular: resembling or relating to twilight
crevice: slit, crack, cleft, gap
critical: very important
critically endangered: facing an extremely high risk of extinction in the wild
critter: informal word for any living creature
crustacean/s: mainly aquatic arthropod typically with body covered with a hard shell or crust, such as lobsters, shrimps, crabs, barnacles, wood lice
cryptic: secret, mysterious, hidden
ctenoid scales: scales with spines or comb-like teeth along their free edge
cull/culling: reduce the population of (a wild animal) by selective slaughter
culprits: wrongdoer, offenders
cultivate: breed, raise, tend to, plant
cultural significance: of some value for past, present or future generations
culvert/s: tunnel carrying a stream or open drain under a road or railway
currency: money, legal tender
cyanobacteria: formerly known as blue-green algae, are photosynthetic microscopic organisms that are technically bacteria
cylinder, cylindrical: 3D shape like a tube with circles on either end
debris: **scattered** pieces of **rubbish** or remains
decade: 10 years
decline/declining: drop, gall, decay, weakening
decorticate/decorticating: remove the **bark**, **rind**, or **husk** from.
deep (anatomy): further from the surface
defence mechanism/s: adaptations that help avoid predators or mitigate adverse effects.
defense, defensive: protecting or defending against something
deforestation: the action of clearing a wide area of trees
degrade, degradation: reduce in quality, lessen, weakening, deterioration
dense: thick, heavy, impenetrable, solid
derivation (words): where the word/s came from
destruction: breaking something apart, causing so much damage it can't be used or fixed, or even removed
detritus: small pieces of disintegrated material such as organic matter
difference between endemic and native? An endemic species is a native species but found only in a particular area, large or small
digest: to break down food into smaller pieces to use for energy and growth
digestive system: the parts of the body that work together to turn food and liquids into the building blocks and fuel that the body needs for energy and growth
digit/s: finger, thumb or toe
dimorphic: two different forms
dioecious: having the male and female reproductive organs in separate individuals
disease: illness or sickness characterised by specific signs or symptoms
dissipate: disappear, waste or fritter away
distal (anatomy): further from the centre
distinction/s: noticing the difference between two things
distinguish, distinguishes: mark off as different
distinctive, distinctively: having a special quality, style, colour, etcetera
distribute, distribution: divide into, disperse through a space or over an area, spread, scatter
diurnal: of or during the day.
diverse/diversity: different or having variety
dominant: more strong than others
dorsal, dorsum (anatomy): of, relating to, or situated at the back, or dorsum of
drainage: removing excess water from an area

drupe: a type of fleshy fruit that contains a single hard seed enclosed in a stony layer
duck blind: shelter, often camouflaged with reeds and grasses, for hidden observation
dusk: time of day when the sun is setting and it's starting to get dark, just before nightfall
dusky (of colour): of a darker colour, shadowy
dwindle: become less or smaller gradually
echolocation/echolocate: the location of objects by reflected sound
ecosystem: a geographic area where plants, animals, and other organisms, as well as weather and landscapes, work together to form a bubble of life
ectomycorrhizal: a symbiotic association of fungi with the feeder roots of higher plants
edible: fit to be eaten as food, eatable
eel: long snake-like fish
elaborate: complicated, detailed, **complex**
elusive: tricky, difficult to catch, shy
endemic: native species found only in a particular area
embryo: baby in its very early stages, before it's born; animal or plant that's just starting to grow and develop
endangered: there is a very high risk of extinction due to rapid population declines of 50 to more than 70% over the previous 10 years, a current population size of fewer than 250 individuals
environment, environmental: the natural surroundings that make up the world we live in
EPBC: Environment Protection and The Biodiversity Conservation Act 1999 (EPBC Act) is Australia's main environmental law
epicormic buds: dormant buds that lie beneath the bark of a tree
epipelagic (ocean zone): sunlight zone
estimate, estimated, estimating: guess, projection
estuary/estuarine: an area where a freshwater river or stream meets the ocean
eukaryotes: any cell or organism that possesses a clearly defined **nucleus**
European settlement: 1788-1810, early European settlement
evade: avoid or get away from something, often by being clever or sneaky
evolve, evolved: grow, mature, advance
evolution, evolutionary: change, expand, progress, transform
excavate: dig out, burrow, uncover
excretion/s: discharge by excreting
exotic: of foreign origin or character, introduced from abroad
extensive, extensively: covering a large area
extinct/extinction: having died out, no longer existing
extreme weather events: occurrences of unusually severe weather or climate conditions that can cause devastating impacts on communities and agricultural and natural ecosystems
exudate: substance coming out, ooze
fatal: causing death
faunal region: any of six or seven areas of the world defined by animal geographers on the basis of their distinctive animal life
feral: wild, having reverted to a wild state
fertilised, fertilisation: make productive, capable of development
fire regimen: pattern, frequency, and intensity of the bushfires and wildfires that prevail in an area over long periods of time
flank/s: side between ribs and hips
flock: group of animals, especially birds like geese or chickens, or sheep and goats, that stay together

follicles: a dry fruit that splits open along a suture to release seeds

forage, foraging: look for food

formidable: something that is really strong, impressive, or even a little scary, but also something you might respect

fossil: the preserved remains or traces of a living thing that lived long ago

fragment, fragmented, fragmentation: broken off, scattered, isolated, separated

frugivore/frugivorous: an animal that thrives mostly on raw fruits or succulent fruit-like produce of plants such as roots, shoots, nuts and seeds

fungus, fungi, fungal: simple organism, or living a thing that is neither a plant nor an animal

gangly: tall and thin and moving with a loose-jointed awkwardness

generalist feeder: feeds on a wide variety of things and thrive in various environments

generic: referring to all members of a genus

gene/s: the basic unit of heredity. This means that genes determine what traits are passed down from a female and a male to their offspring. Genes exist inside the cells that make up living things

genetic/s: traits carried in genes, which are like special instructions for making a body

genus: the major division of a family, usually of more than one species

germinate/germination: when a seed starts to grow into a plant

gestation: the process or period of developing inside the **womb** between **conception** and birth

glabrous: smooth and hairless

global/globally: worldwide, comprehensive

globose: having the shape of a globe or sphere; round

glossy: shiny, lustrous, polished

graffiti: writing or drawings, **scribbled**, **scratched**, or sprayed, usually illicitly

granivore/granivorous: an animal that primarily feeds on seeds and grains

granular: something that is made of tiny grains or particles, like sand or sugar, or that looks like or feels like it is made of those small bits

graze, grazing: eating, feeding, nibbling, munching

groundwater: water below surface of ground

habitat/s: place where usually found, natural environment of an organism

harem: a group of female animals sharing a single mate

harvest, harvested: gather food

hatch, hatching: baby animal coming out of its egg, or that an egg is breaking open to release the baby

hatchling: young bird, reptile or fish just hatched from egg

heath/heathland: an area of open uncultivated land, typically on acid sandy soil, with characteristic vegetation of low-growing plants, shrubs and grasses.

herbicide/s: a substance that is **toxic** to plants

herbivore/herbivorous: an animal that primarily eats plants

herper/s: a person who is interested in reptiles and amphibians, or herps

hibernate, hibernation: spend the winter in a **dormant** state

hinder: make harder or slower to do, make difficult

honeydew: sugary material secreted by some insects such as aphids, scale

horizontal: parallel to horizon

host plants, host trees: plants that an organism lives on and lives off (parasitic or semi-parasitic plants)

hostile: aggressive, warlike

human disturbance: people disturbing the environment: for example - tree removal. road building construction near shorelines and streambanks

humid, humidity: moisture or water vapour in air

hydrology: the branch of science concerned with the properties of the earth's water, and especially its movement in relation to land.

iconic: representative of

illegal: against the law, not legal

immature: something that is not fully developed or grown

immune to: not being affected by something

impoundment (of streams): damming, enclosing

incisor/s: a narrow-edged tooth at the front of the mouth, adapted for cutting

incubate, incubated, incubation: brooding

indigenous: having originated in and being produced, growing, living, or occurring naturally in a particular region or environment

indigenous people: First Nations people

inferior (anatomy): below, lower than

inflorescences: the complete flower head of a plant including **stems**, **stalks**, **bracts**, and flowers

inhabit, inhabits: live in, occupy, populate, reside

insecticide: an agent (usually chemical) that destroys insects

insectivore/insectivorous: a carnivorous animal or plant which eats insects

instar: a developmental stage between moults, or shedding of the exoskeleton

intertidal: between high and low water marks

intimidate/d: scare, frighten, bully, terrify

introduce/introduced (of species): plant or animal that wasn't originally there but was brought in by humans, either on purpose or by accident

invasive: threatening a plant or animal community, by spreading prolifically

invertebrate/s: animals without backbones

iridescent: something that changes colours when you look at it from different angles

iridophores: an iridescent chromatophore

iris (of eye): the coloured tissue at the front of the eye that contains the pupil in the centre

isolation, isolated, isolate: detach, separate, place apart

IUCN: International Union for the Conservation of Nature

joey/s: young marsupial animal

juvenile/s: youth, junior, infant, young

keystone species: an organism that has a disproportionately large impact on its ecosystem relative to its abundance

kinetoplast, kinetoplastids: a group of flagellated protozoans that are distinguished by the presence of a DNA-containing region, known as a "**kinetoplast**", in their single large mitochondrion

labellum: a central **petal** at the base of an **orchid** flower, typically larger than the other **petals** and of a different shape

lactate, lactating: producing and secreting milk from the mammary glands

lantana thicket: compact shrub which can grow in dense thickets, areas of trees and bushes growing close together

larva, larvae: immature, wingless, feeding stage of an insect that undergoes complete metamorphosis

lateral (anatomy): towards the side

lichen: a complex, plantlike organism formed by a symbiotic relationship between a fungus and a photosynthetic partner, which can be an alga or a cyanobacterium

life cycle: the series of changes in the life of an **organism**

lifespan: the period of time of existence from birth to death or the end of useful function

lignotuber: a woody swelling at the base of a plant that contains dormant buds, nutrients, and carbohydrates

ligule (botany): a narrow strap-shaped part of a plant, especially a membranous scale on the inner side of the leaf **sheath** at its junction with the blade in most **grasses** and **sedges**

litter: number of young born in one birth

littoral: relating to or situated on the shore of the sea or a lake

littoral rainforest: a closed forest community that grows near the ocean, usually within 2km

local, locally, localised: restricted to one place

locally extinct: no longer present in a specific geographic region but still exists elsewhere

longitudinal/longitudinally: running **lengthwise** rather than across.

macro- (prefix): large; large-scale.

mammal: warm-blooded vertebrate having hair or fur, producing milk for the young and giving birth to live young – exceptions are the **monotremes.**

Māori people: indigenous Polynesian people of mainland New Zealand

marine: of or relating to the sea

marsh/marshes/marshland: type of wetland ecosystem characterized by poorly drained mineral soils and by plant life dominated by grasses

marsupial: animals in the order *Marsupialia* which have a pouch for their young

marsupium: pouch found in female marsupials

maternity: the period during pregnancy and shortly after childbirth

maternity colonies: group of pregnant or lactating female bats that gather together in a roost to give birth, nurse, and raise their young

mature, maturity: fully developed, having completed natural growth and development

mechanism: the fundamental processes involved in or responsible for an action, reaction, or other natural phenomenon

medial (anatomy): towards the middle

melody/melodious: sounding pretty and musical

membrane/s: sheath, sheet, film of tissue

mesial: being or located in the middle or a median part

mesopelagic (ocean zone): twilight zone

metabolism: the chemical reactions in cells that change food into energy

metamorphosis: change in life cycle e.g. pupa to caterpillar to butterfly

methodical: orderly, structured, deliberate

micro- (prefix): very small, tiny

microalgae: microscopic algae invisible to the naked eye; phytoplankton typically found in freshwater and marine systems, living in both the water column and sediment. They are unicellular species which exist individually, or in chains or groups

microendemic: species with very small geographic distributions (ranges)

microorganism: a class of biological cells constituting a vast domain of **prokaryotes**, and typically include bacteria, fungi, viruses, and **microalgae**, etc. **Microorganism** is micrometres in size and possesses various shapes ranging from rod, spheres and spirals

microphyte: a microscopic plant

microscopic: incredibly small, unable to be seen without a microscope

migrant: drifter, traveller, wanderer

migrate, migration, migrating: going from one country, region or place to another

migratory: characterised by migration

mimic, mimicking, mimickry: copy, simulate, imitate

mitochondria, mitochondrium: the tiny powerhouses inside our cells! They break down the food we eat to give us energy, so we can move, think, and do everything else we do...think of them as little energy factories

moist: slightly wet, damp

moisture: the presence of a liquid, especially water, often in small amounts

molar: the back teeth in the mouth, shaped like flat, grinding surfaces

mollusc/s: animal that has a soft body, no spine, and is often covered with a shell

monitor: observe, watch, keep an eye on

monoecious: having both the male and female reproductive organs in the same individual

monogamous: having only one mate at a time

monopolise: have complete and sole control over, take over control

monotreme/s: mammals who lay eggs and have no teats but secrete milk to feed their young

morphology, morphological: form of a thing

motivate. motivated: inspire, persuade, prompt, spark

moult: shed, cast, peel - when an animal sheds its old outer covering (skin/feathers, for example) to make way for a new one or new ones

myth, mythology: belief, legend, lore, tradition

native: an animal or plant that has lived in a specific area for a long time, naturally, without being brought in from somewhere else

navigate, navigation: find a route to a destination

navigator: a person or animal who navigates

NCA: Nature Conservation Act 1992

nectar: sweet, sticky liquid made by flowers

nectarivore, nectarivorous: an animal which derives its energy and nutrient requirements from a diet consisting mainly or exclusively of the sugar-rich nectar produced by flowering plants

nocturnal: active at night, relating to night

nomadic: wandering, travelling, roving

no-take: not allowed to be taken/caught

nucleus/nuclei: the structure in a cell that contains the chromosomes

nutrient/s: source of nourishment, food

nutrient-rich: rich in nutrients, nourishing food

nymph (insect): young insect that looks a bit like the adult, but smaller and without wings

oceanic: marine, of the oceans

Olongburra: This specific name refers to the aboriginal people, the Olongburra, who once lived in the northern half of K'gari (Fraser Island)

omnivore, omnivorous: eating both animal and plant foods

opportunistic: taking opportunities when they present

opposable, opposing: able to be placed opposite to something else

organelle: a **subcellular** structure that has one or more specific jobs to perform in the cell, much like an organ does in the body. Among the more important cell **organelles** are the **nuclei**, which store **genetic** information; **mitochondria**, which produce chemical energy; and **ribosomes**, which assemble proteins

organic: live, living, natural

organic material: matter that is made of organic compounds from the remains of organisms such as plants and animals, and their waste products

organism: animal, plant, structure, creature

ovary: the organ in a female animal that produces eggs and certain hormones; the part of a flower that contains the seeds and grows into a fruit

overexploit, overexploitation: overuse, excessive use of resources

overfish, overfished, overfishing: catching too many fish at once so that the breeding population is too reduced to be able to recover

overhunt, overhunted, overhunting: the act of hunting animals or birds to an excessive degree, which can be harmful to the population or the environment

overgraze, overgrazed, overgrazing: allowing grazing animals to reduce grazing land to beyond the point of recovery

ovum: an egg cell

ovule: an outgrowth of the ovary of a seed plant that after fertilization develops into a seed

paralysis: inability to move, can sometimes affect entire body systems such as breathing, movement, heart beat

parasite, parasitic: an organism that lives on or in an organism of another species (the host) from the body of which it obtains nutriment

passerine: relating to or denoting birds of a large order distinguished by having feet that are adapted for **perching**, including all **songbirds**

pastural/pastoralism: something related to rural life, especially the life of shepherds or country life

patagium/patagia *(pl.)***:** wing membrane (bat), extensible fold of skin (insects, gliding mammals and reptiles), small process on the anterior thorax seen especially among butterflies and moths

pectoral fin: see fish anatomy diagram

peduncle (botany): the stalk of a flower

pelvic: having to do with the pelvis

pelvices: plural of pelvis

pelvic fin: see fish anatomy diagram

pelvis: the bony structure of the body below the abdomen that is located between the hip bones and contains the bladder and rectum

perch, perching: place where a bird can sit and rest, like a branch or a stick

perennial: lasting or existing for a long or apparently infinite time; (of a plant) living for several years

perimeter/s: boundary, edge, borderline

pesticide/s: a substance used for **destroying** insects or other **organisms** harmful to **cultivated** plants or to animals

photosynthesis: how plants make their own food using sunlight, water, and air to produce oxygen and sugar, which they use for energy

phytoplankton: consisting of **microscopic** plants

piscivore/piscivorous: a carnivorous animal that primarily eats fish

pistil: the female reproductive part of a flower

placid: gentle, peaceful, quiet

plankton, planktonic: floating, drifting, or somewhat motile organisms occurring in a body of water, primarily comprising microscopic algae and protozoa

plastron: the ventral part of a turtle shell turtle

plume/plumage: feather (singular), feathers of a bird, collectively

poison, poisonous: likely to cause serious harm or death
pollen: a fine powdery substance, typically yellow, consisting of microscopic grains discharged from the male part of a flower or from a male cone. Each grain contains a male **gamete** that can **fertilise** the female **ovule**, to which pollen is transported by the wind, insects, or other animals
pollination: the act of transferring pollen grains from the male anther of a flower to the female stigma to help make new plants
pollinator, pollinators: animal, insect or other agent that transfers pollen to a plant allowing fertilisation
pollution: the introduction of harmful materials into the environment
polychaete worm: poly means 'many', chaete means 'hair', so these are a species of marine worm, commonly known as bristleworms.
polyclad, polyclads: any free-swimming, marine flatworm of the order Polycladida, having a broad, flat body and a many-branched gastrovascular cavity
polyoestrous: can raise several litters a year
population: number of inhabitants of an area
posterior (anatomy): behind, towards the back
precarious: dependent on circumstance, unsure, indeterminate, unreliable, doubtful
precipitation: what happens when water in the sky (clouds) falls back down to Earth as rain, snow, sleet, or hail
precocial: (of a young bird or other animal) **hatched** or born in an advanced state and able to feed itself and move independently almost immediately
predator/s, predation: an organism that exists by preying on other organisms
predict, prediction: to be able to tell in advance
pregnant: containing a developing embryo, foetus, or unborn offspring within the body
prehensile: adapted for seizing, grasping, or taking hold of something
premature: happening or done too soon, or before the expected or natural time
prey, preying: an animal, insect or organism hunted or taken for food, especially by a carnivorous animal
primary: most important, main
prokaryotes: organisms whose cells lack a **nucleus** and other **organelles**. **Prokaryotes** are divided into two distinct groups: the **bacteria** and the **archaea**, which scientists believe have unique evolutionary lineages. Most **prokaryotes** are small, single-celled organisms that have a relatively simple structure
probe/probing: examining with the beak
prominent: standing out, obvious
protozoa, protozoan: any of a diverse group of eukaryotes that are primarily unicellular, existing singly or aggregating in colonies, are usually non-photosynthetic and often classified further into phyla according to their capacity for means of motility
protuberance/s: a rounded, outward projection or swelling on a surface
proximal (anatomy): closer to (root of the limb)
puggle: baby echidna or platypus
pulmonate, pulmonates: having lungs or lunglike organs
pupa, pupae *(Pl.)***:** an insect in the nonfeeding, usually immobile, transformation stage between the larva and the adult stage
pupil (eye): the black opening in the middle of the coloured part of your eye
pursue: chase, track, follow, trail, hunt
pursuit: chase, hunt, quest, run

quarry (prey): something being hunted or pursued, specifically prey

quarry, quarrying: an open excavation usually for obtaining building stone, slate, or limestone

radio-tracking: a way of sending sounds or other information through the air from a transmitter attached to the animal to enable researchers to track it

rainforest canopy: the denser layer of greenery about 20 feet thick formed by the next tallest trees.

range (of species): an area where a member of a particular species can be found during its lifetime

rapid/rapidly: quickly, quick

raptor: usually a bird or prey, such as an eagle or a hawk

rare: not common or usual

recognise, recognition: seeing something you have seen before or can clearly identify by other means

recreation, recreational: fun activity, amusement, hobby

refuge: place that offers safety and protection from danger, trouble, or bad weather

regenerate: the process of restoring damaged or missing cells, tissues, organs, or even entire parts

remnant: remains, fragment, piece

reproduction/reproduce/reproductive: the production of **offspring**

reptile: cold-blooded vertebrate

residential development: housing development

respiratory system: breathing system, lungs

restrict, restricted: confine, restrain, regulate, limit

retina, retinae *(Pl.)***:** inside back of eye of several layers

revere, revered: respect or admire greatly

reverse: backward, contrasting, opposite, contrary, change

reversible (physiological): something that can be changed from back to front or front to back

rhomboid, rhomboidal: parallelogram in which adjacent sides are not equal

ribosome: an intercellular structure made of both RNA and protein, and it is the site of protein synthesis in the cell

riparian: situated on the bank of water

robust: strong and healthy

rodent/s: a **gnawing** mammal of an order that includes rats, mice, squirrels, hamsters, **porcupines**, and their relatives, distinguished by strong constantly growing incisors and no **canine** teeth

rodenticide: a type of poison used to kill or control rodents

rods (eyes): specialised cells which are sensitive photoreceptors to light and help us see in low-light conditions. They are primarily responsible for night vision and peripheral vision, but they don't perceive colour

roost, roosting: perch where birds rest or sit

rufous: tinged with red, brownish red

rump: hindquarters, buttocks, sacral region of a bird

rural: relating to the countryside and not to towns or cities

sagittal (anatomy): dividing body into right and left

saline: containing or **impregnated** with salt

salinity: saltiness

samara: a dry indehiscent usually one-seeded winged fruit

satellite: an artificial body placed in **orbit** round the earth or moon or another planet in order to collect information or for communication.

satellite tracking: using satellites orbiting earth to track

satellite transmitter: device which sends a signal to a satellite
saturate, saturated: soaked, soggy, waterlogged
savannah: flat grassland, plain
scarce, scarcity: not plentiful, hard to find or hard to get
school, schooling (of fish): group of fish moving together
sclerophyll forests : vegetation that has adapted to long periods of dry, hot weather
sea squirt: invertebrate, usually fixed potato-shaped organism using sac-like filter feeding
secretive: keeping secrets
sedentary: staying in one place, not moving much
sedge: plant that looks like grass, but it grows in wet areas
sediment, sedimentation: organic matter settling to bottom of a liquid
sedimentary: of or relating to rocks formed by the deposition of sediment
segment/s: part of, divided parts
semi- (prefix): half, partially, incompletely
semi-parasitic: plant which gets water or food from host but can also produce its own food
semi-prehensile: able to grasp but cannot really hang on tightly
semi-transparent: almost see through
sepal: the green, leaf-like part of a flower that protects the bud before it blooms and supports the petals when the flower is open
serrated: having a jagged or toothed edge, like a saw
settlement: a place, typically one which has previously been **uninhabited**, where people establish a community
shellfish: seafood like molluscs, crustaceans
sheoak: type of tree
sibling: each of two or more children or offspring having one or both parents in common; a brother or sister
sift: an act of sifting something, especially so as to **isolate** that which is most important
significant, significance: very important or has a special meaning
silt: fine sand, clay or other material carried by running water
siltation: water pollution by silt
skim, skimming: to pass lightly or hastily; glide or skip along, above, or near a surface
snout: the long nose and mouth of an animal that sticks out
snout-to-vent: a measurement of the body length of an animal, specifically from the tip of the snout to the cloacal opening, or vent
social: friendly, like company
solitary: by itself, single
spawn, spawning: reproduce, generate (usually fish or frog eggs)
sparse: something that is thinly spread out or not dense
species: one of the classes of things included with other classes in a genus
specific: clear-cut, particular, precise, unique
specimen: sample, example
spinifex: grass plant with stiff, sharp-pointed leaves that grow in dense masses
spinose: having many sharp, pointed things sticking out; spiny, prickly
spire: a part of the coiled shell of molluscs. The spire consists of all of the whorls except for the body whorl
spoil heap: an untidy pile or mass usually of refuse material from an excavation
spore: a reproductive cell that can develop into a new individual without uniting with another reproductive cell
stalling: stopping (flying)

stance: way of standing physically or figuratively

stamen: the male part of a flower that makes pollen

stigma/s: the part of a flower that receives pollen and is the tip of the **pistil**

stolon (botany): a **creeping** horizontal plant stem or runner that takes root at points along its length to form new plants

stochastic: random or involving chance

storm front: border between two different kinds of air, often with different temperatures; when a front is created, weather such as storms and **precipitation** happen

streamers (tail feathers): tail streamers on birds are the elongated narrow tips of the tail feathers; elongated, specialised tail feathers that are often used for display and communication, particularly during mating rituals. They can be quite long, sometimes exceeding the bird's body length, and are often vibrant in colour or pattern

streamline, streamlined: designed to have the least resistance

sub- (prefix): underneath, below, less than

subcellular: of less than cellular scope or level of organisation

subpopulation: specific small section within a larger group

subspecies: additional species within a grouping

subtropical: bordering on the tropics; nearly tropical. pertaining to or occurring in a region between tropical and temperate; semitropical

superficial (anatomy): on the surface

superior (anatomy): above, higher than

supra- (prefix): above, beyond, or earlier in time

supralittoral: area above shoreline, likely damp

survive, survival: stay alive, staying alive

susceptibility: easily affected, harmed, or influenced by something

swamp: an area of low-lying, uncultivated ground where water collects; a bog or marsh

swamp drainage: emptying water of swamp

swim bladder: special organ inside a fish that helps it float and control its depth in the water

symbiotic, symbiosis: living together in a mutually beneficial way

tail streamers: tail streamers on birds are the elongated narrow tips of the tail feathers; elongated, specialised tail feathers that are often used for display and communication, particularly during mating rituals. They can be quite long, sometimes exceeding the bird's body length, and are often vibrant in colour or pattern

tannin: astringent vegetatation compound, reddish/brown

tassel, tasselled: fringed lobes of skin similar to tassels which are made of multiple cords, usually hanging down

tawny: dark or dull yellowish-brown

temperate: moderate in respect to temperature

termitarium/termitaria: termite nest/s

terrestrial: land based, growing or living on land

territory, territories, territorial: a tract of land, somewhere that the animal recognises as their territory which they will protect if they are territorial

thermal: using, resulting from, or producing heat

thermal current: energy that is transferred from one body to another because of a difference in temperature creating a current

thorax: the portion of the body between the head and the abdomen

threat: alarm, danger, hazard, menace, peril, risk, warning

threatened: in danger or likely to be harmed,

or to become more significantly vulnerable or endangered

tissue: a group of similar cells and their intercellular material that work together to perform a function

tolerate: put up with

torpor: a state of physical or mental inactivity; lethargy.

totem, totem animal: a natural object or animal that is believed by a particular society to have spiritual significance and that is adopted by it as an emblem

toxic: poisonous

trample. trampled: when hard-hooved animals stamp or tread heavily on the ground

transmit, transmitted: to send or pass something from one place or person to another, like a message or a disease

transparent: see-through, light passes through completely like clear glass

transverse (anatomy): dividing body into upper and lower sections

trauma, traumatic: frightening, dangerous, or violent event that poses a threat to life or bodily integrity; wound or injury which can be both physical and/or psychological, or both

trawl: (myotis) fish by dragging feet through water

tree canopy: the upper parts of a tree, leaning into neighbouring trees

trenches (ocean zone): hadal zone

tropics, tropical: regions of Earth closest to the imaginary line called the equator, which runs around the middle of the planet. It's always warm in the tropics, with high humidity and plenty of rain

truffle: a fungus that grows underground and is edible

truncate: to cut off short; abbreviate, cut, foreshorten, shorten

trypanosome: genus of **kinetoplastids**: a monophyletic group of unicellular parasitic flagellate protozoa

tuber/s: underground part of a stem or rhizome, like potatoes or sweet potatoes.

tubercles, tuberculate: a little bump or knob, kind of like a tiny wart, that can grow on plants or on the body, especially on bones

tundra: major zone of treeless level or above the timberline on high mountains, rolling ground found in cold regions, mostly north of the Arctic Circle

turbulent/turbulence: characterised by conflict, disorder, or rapid, uneven changes. It can describe situations, periods of time, or even physical movements like air or water flows. Think of it as the opposite of smooth, stable, or calm

tympanum: the eardrum or tympanic membrane

understorey: the layer of plants and especially the trees and shrubs between the forest canopy and the ground cover

unicellular: consisting of a single cell

unique, uniqueness: one of a kind, unlike anything else

urban: relating to a particular place, town, village, in this instance development due to humans needing to have more homes

vagrant: wandering from place to place

vary, variable, variation: the quality or state of being different or diverse, not uniform

vegetarian: eating only plant matter

vegetation: plants

vehicular traffic: movement of cars, trucks, and other vehicles on roads

venom, venomous: poison, poisonous

venom glands: the specialised glands

responsible for producing the poisonous secretions of an animal, often associated with spines, teeth, stings, or other piercing devices

vent (anatomy): the external opening of the **cloaca**, a single posterior orifice that serves as the only opening for the digestive, urinary, and reproductive tracts in many animals, including birds, reptiles, amphibians, some fish and mammals

ventral (anatomy): towards the front

ventrolateral/ventrolaterally: the front and side of a body or body part

verify, verified, verification: prove the truth of, evidence

vertebra, vertebrae, vertebrate: spinal column or backbone

vertical: upright, top to bottom

vicinity: the area around something

virus: an infectious microbe consisting of a segment of nucleic acid (either DNA or RNA) surrounded by a protein coat. A virus cannot replicate alone; instead, it must infect cells and use components of the host cell to make copies of itself

vivid: strikingly bright or intense

vulnerable: weak, unsafe, defenceless, sensitive

water hole: a natural puddle or small lake in a dry area where animals go to drink

water-logged: saturate with water

water table: the boundary between saturated and unsaturated ground

weed: plant growing where it's not wanted

weir/s: bank, barrier, ditch, dike

wet sclerophyll forest: characterised by very tall eucalypt trees usually in moist cloudy uplands with up to 2500mm of rainfall each year

wetlands: an area of land that's covered by water or has waterlogged soil, meaning the ground is full of water, like a sponge

whorl/s: something that swirls or winds around like a spiral

wingspan: the maximum extent across the wings of a bird or other flying animal, measured from tip to tip

xanthophore/s: a **chromatophore** containing a yellow pigment that is typically a **carotinoid** and occurring especially in fishes and crustaceans

REFERENCES AND CITATIONS:

1. IUCN (2024, July 19). *IUCN Red List*. International Union for the Conservation of Nature. Retrieved August 8, 2024, from https://www.iucnredlist.org/
2. Brigalow Belt - Wikipedia. https://en.wikipedia.org/wiki/Brigalow_Belt
3. https://wetlandinfo.des.qld.gov.au/wetlands/facts-maps/bioregion-brigalow-belt-brb/
4. Paradise parrot - Wikipedia. https://en.wikipedia.org/wiki/Paradise_parrot
5. Discover 16 Extinct Types Of Parrot - Pets Tutorial. https://petstutorial.com/discover-16-extinct-types-of-parrot/
6. Researchers Are Elated To Find Extremely Rare Polka-Dotted Babies Wriggling In Their Mom's Pouches. https://aubtu.biz/extremely-rare-polka-dotted-babies
7. What does a chuditch look like? – Quick-Advices. https://quick-advices.com/what-does-a-chuditch-look-like/
8. On the Trail of the Plains-wanderer - Reading Time. https://readingtime.com.au/on-the-trail-of-the-plains-wanderer/
9. The Plains Wanderer Project. https://apdha.org.au/the-plains-wanderer-project
10. The bulloak jewel butterfly and how the web of life affects us all — from mistletoe to humans - ABC News. https://www.abc.net.au/news/science/2019-01-06/bulloak-jewel-butterfly-ant-mistletoe-a-web-woven-across-species/10477430
11. Gulbaru Gecko | Phyllurus gulbaru. https://www.edgeofexistence.org/species/phyllurus-gulbaru/
12. Conservation of Imperiled Crayfish—Euastacus jagara (Decapoda: Parastacidae), a Highland Crayfish from the Main Range, South-Eastern Queensland, Australia. Journal of Crustacean Biology. https://doi.org/10.1651/09-3207.1
13. Euastacus - Wikipedia. https://en.wikipedia.org/wiki/Euastacus
14. Hipposideros semoni - Living Wonders. https://livingwonders.org.au/evidence/hipposideros-semoni/
15. Bridled nail-tail wallaby - Wikipedia. https://en.wikipedia.org/wiki/Bridled_nail-tail_wallaby
16. 5 Fun facts about the silver-headed antechinus – WWF-Australia | 5 Fun facts about the silver-headed antechinus | WWF Australia. https://wwf.org.au/blogs/5-fun-facts-about-the-silver-headed-antechinus/
17. Gouldian Finch / Bird Species / Birdwatching | Ornithology Education. https://www.ornithology.org/birdwatching/bird-species/gouldian-finch
18. . https://www.psittaciformes.com/species/cacatuidae/lophochroa.html
19. Lophochroa leadbeateri. https://www.scientificlib.com/en/Biology/Animalia/Chordata/Aves/LophochroaLeadbeateri01.html
20. 20 Fun Koala Facts for Kids to Spark Classroom Conversation | Teach Starter. https://www.teachstarter.com/us/blog/koala-facts-for-kids/
21. Black-throated finch - Wikipedia. https://en.wikipedia.org/wiki/Black-throated_finch
22. Albert's lyrebird - Wikipedia. https://en.wikipedia.org/wiki/Albert's_lyrebird
23. Boggomoss Snail | Facts About Snails. https://factsaboutsnails.com/types-of-snails/native-australian-snails/helicoid-snail-camaenidae/boggomoss-snail/
24. Australian Masked-owl | The Peregrine Fund. https://www.peregrinefund.org/explore-raptors-species/owls/australian-masked-owl
25. Eungella Honeyeater - The Australian Museum. https://australian.museum/learn/collections/natural-science/ornithology/ornithology-collection-eungella-honeyeater/
26. The jewel-eyed gecko with a foul-smelling secret - Australian Geographic. https://www.australiangeographic.com.au/blogs/creatura-blog/2017/12/the-jewel-eyed-gecko-with-a-foul-smelling-secret/
27. Cascade Tree Frog articles - Encyclopedia of Life. http://media.eol.org/pages/52569957/articles
28. February 18: White-bellied Sea Eagle - daily-article-l - lists.wikimedia.org. https://lists.wikimedia.org/hyperkitty/list/daily-article-l@lists.wikimedia.org/thread/G6QWJW5PMOHELJ3YGCJVXQYCZEV5SU3H/
29. Osprey Overview, All About Birds, Cornell Lab of Ornithology. https://blog.allaboutbirds.org/guide/Osprey/overview
30. Yellow chat - Wikipedia. https://en.wikipedia.org/wiki/Yellow_chat
31. Blue-winged Parrot - BirdLife Australia. https://birdlife.org.au/bird-profiles/blue-winged-parrot/
32. Star finch - Wikipedia. https://en.wikipedia.org/wiki/Star_finch
33. White-throated needletail - Wikipedia. https://en.wikipedia.org/wiki/White-throated_needletail
34. White-throated Needletail | BIRDS in BACKYARDS. https://www.birdsinbackyards.net/species/Hirundapus-caudacutus
35. Stagonopleura guttata. https://www.scientificlib.com/en/Biology/Animalia/Chordata/Aves/StagonopleuraGuttata01.html
36. About - Alaska EXCEL. https://alaskaexcel.org/about/
37. Phyllurus amnicola - Wikipedia. https://en.wikipedia.org/wiki/Phyllurus_amnicola
38. Northern bettong - Wikipedia. https://en.wikipedia.org/wiki/Northern_bettong

To claim your Conservation Club reward badge for THE BRIGALOW BELT environment:

- **Answer the following 20 questions.**
- **Send either a photograph or a scan of one colouring page** which you have coloured in reasonably correctly as far as the colours of your chosen species are concerned!

*** Postage will apply both in Australia and overseas. This will be estimated and sent to you for approval via email prior to shipping.**

Let's see what you have learned, whether you are 8, 80, or somewhere in between!

The answers can be found throughout the book.

1. How many species can you find in the book that are VULNERABLE on any of the IUCN, EPBC or NCA lists?
2. Which parrot is, sadly, extinct?
3. Which little bird lives on the bare saltbush plains, grasslands and herb fields in three Australian states?
4. Which treefrog lives in only one national park southwest of Gladstone?
5. Which mammals represent the second biggest group of mammals in the world?
6. Which marsupial was declared extinct in 1937 and then rediscovered?
7. Which gliding possum is the largest in eastern Australia and is very furry?
8. Which bird's chicks have luminous nodules inside their mouths to direct their parents to their hungry mouths?
9. Which burrowing animal has cube-shaped poo?
10. Which carnivorous bird is the second-largest nocturnal raptor in Australia?
11. Which native shrub is a host plant for red-eye butterflies?
12. Which honeyeater was one of the last new species of birds discovered in Australia?
13. Which reptile squirts foul-smelling, flammable liquid from its tail as a defence mechanism?
14. Which bird's plumage looks so much like the bark of trees that it is extremely hard to find and described as a master of disguise?
15. Which species of native fish has recently been discovered in Dismal Swamp?
16. Which plant smells like mice?
17. Which subspecies of small bird is believed to number less than 50 individuals and may possibly be extinct?
18. Which species of bird flies almost twice as fast as a cheetah can run?
19. Which plant is considered a living fossil from the Jurassic period?
20. Which fierce little marsupial is sometimes called a native cat?

FROG ID – Things to look for when trying to identify frogs….

To be able to accurately identify frogs, you will need to develop some knowledge of their physical characteristics far beyond the information given here. These are just some very basic clues and are based on information from the book listed below, which, if you are a keen **herper** or likely to become one, I cannot recommend highly enough! This book narrows things down using a dichotomous key which you can go through step-by-step!

> **Sanders MG (2021) *Photographic Field Guide to Australian Frogs.* CSIRO Publishing, Melbourne.**

Head shapes:

> ***Dorsal outline:*** rounded, pointed, truncate
>
> ***Profile outline:*** rounded, sloping, bent, truncate, acute

Canthus shapes: rounded, angular, curved, and nearly straight

Pupil shapes: vertical, round, rhomboidal, horizontal

Skin textures: smooth, coarsely granular, finely granular, finaly granular with prominent tubercles and rounded ridges, irregularly textured, irregularly tuberculate, spinose and large rounded glands

Skin colour patterns: spots, blotches, ocelli, flecks, marbling, streaks and stripes, bands or bars

> **NB:** Sometimes temperatures or the substrate on which they are resting can influence the colours that they display – just to complicate things!

Frogs have four fingers and five toes!

> Some have long, skinny digits, others have shorter, stubby digits.
>
> Some have digits with terminal tips or swelling or 'pads', others do not have terminal digit swelling.
>
> Some have very distinctive webbing, others not so much - there is also variation in where the webbing occurs.

Location/distribution:

> A very good clue as to what many species would be and where they would occur is where they are found. This is beyond the scope of the information I can offer here.

I hope that these physical characteristics you can look for will help you narrow down the species you might have seen. Make sure to take notes of where you were (geographically and in terms of environments), what time of the year it was (seasons and temperatures could be important), and any other observations that you can note down in terms of descriptions. Take photos if you are able to!

- **ABOUT THE AUTHOR**
-
- Meet Elysia Rainwood, an artist and writer with a passion for conservation and environmental protection, particularly the preservation of the diverse animals inhabiting our wonderful Australian continent.
-
- Growing up amongst the rustling leaves of eucalyptus trees, Elysia developed a passion for the unique creatures and the diversity of plant life found in Australia. Elysia always had an eye for detail, being very aware of such things as seasonal changes in the various environments observed. From the depths of the lush, green rainforest to the sun-kissed deserts, her childhood adventures fueled a desire to share the stories of the vulnerable and endangered.
-
- Elysia writes non-fiction books and stories that resonate with the spirit of the tiniest of creatures and soar with the majesty of the wedge-tailed eagle. She likes to focus on the less well-known species, which may exist in your backyard, to ensure that they and their habitats are not overlooked.
-
- **<u>Elysia Rainwood's mission is clear:</u>**
- To enchant young minds with the wonders of Australia's wildlife, fostering a generation of conservationists. Through interesting facts, engaging and reasonably accurate illustrations, she hopes to ignite a spark in the hearts of 8 to 14-year-old readers, inspiring them to protect the precious creatures that share our planet.
-
- Along the way, Elysia hopes to also engage with parents, care givers, educators who can pass on this information as well as those who just enjoy the relaxation that colouring careful illustrations can bring.
-
- So, whether you are a youngster or someone who shares this passion for conservation, join Elysia on this informative journey into the magical realms of nature, where words become wings for the endangered and vulnerable to soar.

INDEX

Made in the USA
Monee, IL
05 November 2026

67385498R00090